ASSESSING RISKS TO

Endangered and Threatened Species

FROM

PESTICIDES

Committee on Ecological Risk Assessment under FIFRA and ESA

Board on Environmental Studies and Toxicology

Division on Earth and Life Studies

National Research Council

NATIONAL RESEARCH COUNCIL
OF THE NATIONAL ACADEMIES

THE NATIONAL ACADEMIES PRESS
Washington, D.C.
www.nap.edu

THE NATIONAL ACADEMIES PRESS   500 Fifth Street, NW   Washington, DC 20001

NOTICE: The project that is the subject of this report was approved by the Governing Board of the National Research Council, whose members are drawn from the councils of the National Academy of Sciences, the National Academy of Engineering, and the Institute of Medicine. The members of the committee responsible for the report were chosen for their special competences and with regard for appropriate balance.

This project was supported by Contract EP-C-09-003 between the National Academy of Sciences and the U.S. Environmental Protection Agency. Any opinions, findings, conclusions, or recommendations expressed in this publication are those of the authors and do not necessarily reflect the view of the organizations or agencies that provided support for this project.

International Standard Book Number-13: 978-0-309-28583-4
International Standard Book Number-10: 0-309-28583-6

Additional copies of this report are available from

The National Academies Press
500 Fifth Street, NW
Box 285
Washington, DC 20055

800-624-6242
202-334-3313 (in the Washington metropolitan area)
http://www.nap.edu

Copyright 2013 by the National Academy of Sciences. All rights reserved.

Printed in the United States of America

THE NATIONAL ACADEMIES
Advisers to the Nation on Science, Engineering, and Medicine

The **National Academy of Sciences** is a private, nonprofit, self-perpetuating society of distinguished scholars engaged in scientific and engineering research, dedicated to the furtherance of science and technology and to their use for the general welfare. Upon the authority of the charter granted to it by the Congress in 1863, the Academy has a mandate that requires it to advise the federal government on scientific and technical matters. Dr. Ralph J. Cicerone is president of the National Academy of Sciences.

The **National Academy of Engineering** was established in 1964, under the charter of the National Academy of Sciences, as a parallel organization of outstanding engineers. It is autonomous in its administration and in the selection of its members, sharing with the National Academy of Sciences the responsibility for advising the federal government. The National Academy of Engineering also sponsors engineering programs aimed at meeting national needs, encourages education and research, and recognizes the superior achievements of engineers. Dr. Charles M. Vest is president of the National Academy of Engineering.

The **Institute of Medicine** was established in 1970 by the National Academy of Sciences to secure the services of eminent members of appropriate professions in the examination of policy matters pertaining to the health of the public. The Institute acts under the responsibility given to the National Academy of Sciences by its congressional charter to be an adviser to the federal government and, upon its own initiative, to identify issues of medical care, research, and education. Dr. Harvey V. Fineberg is president of the Institute of Medicine.

The **National Research Council** was organized by the National Academy of Sciences in 1916 to associate the broad community of science and technology with the Academy's purposes of furthering knowledge and advising the federal government. Functioning in accordance with general policies determined by the Academy, the Council has become the principal operating agency of both the National Academy of Sciences and the National Academy of Engineering in providing services to the government, the public, and the scientific and engineering communities. The Council is administered jointly by both Academies and the Institute of Medicine. Dr. Ralph J. Cicerone and Dr. Charles M. Vest are chair and vice chair, respectively, of the National Research Council.

www.national-academies.org

COMMITTEE ON ECOLOGICAL RISK ASSESSMENT UNDER FIFRA AND ESA

Members

JUDITH E. MCDOWELL (*Chair*), Woods Hole Oceanographic Institution, MA
H. RESIT AKCAKAYA, Stony Brook University, NY
MARY JANE ANGELO, University of Florida
PATRICK DURKIN, Syracuse Environmental Research Associates, Inc., NY
ANNE FAIRBROTHER, Exponent, Inc., WA
ERICA FLEISHMAN, University of California, Davis
DANIEL GOODMAN, Montana State University
WILLIAM L. GRAF, University of South Carolina
PHILIP M. GSCHWEND, Massachusetts Institute of Technology
BRUCE K. HOPE, CH2M HILL, OR
GERALD A. LEBLANC, North Carolina State University
THOMAS P. QUINN, University of Washington
NU-MAY RUBY REED (retired), California Environmental Protection Agency

Staff

ELLEN K. MANTUS, Project Codirector
DAVID POLICANSKY, Project Codirector
KERI STOEVER, Research Associate
NORMAN GROSSBLATT, Senior Editor
MIRSADA KARALIC-LONCAREVIC, Manager, Technical Information Center
RADIAH ROSE, Manager, Editorial Projects
CRAIG PHILIP, Senior Program Assistant

Sponsors

NATIONAL OCEANIC AND ATMOSPHERIC ADMINISTRATION
U.S. DEPARTMENT OF AGRICULTURE
U.S. ENVIRONMENTAL PROTECTION AGENCY
U.S. FISH AND WILDLIFE SERVICE

BOARD ON ENVIRONMENTAL STUDIES AND TOXICOLOGY[1]

Members

ROGENE F. HENDERSON (*Chair*), Lovelace Respiratory Research Institute, Albuquerque, NM
PRAVEEN AMAR, Clean Air Task Force, Boston, MA
MICHAEL J. BRADLEY, M.J. Bradley & Associates, Concord, MA
JONATHAN Z. CANNON, University of Virginia, Charlottesville
GAIL CHARNLEY, HealthRisk Strategies, Washington, DC
FRANK W. DAVIS, University of California, Santa Barbara
CHARLES T. DRISCOLL, JR., Syracuse University, New York
LYNN R. GOLDMAN, George Washington University, Washington, DC
LINDA E. GREER, Natural Resources Defense Council, Washington, DC
WILLIAM E. HALPERIN, University of Medicine and Dentistry of New Jersey, Newark
STEVEN P. HAMBURG, Environmental Defense Fund, New York, NY
ROBERT A. HIATT, University of California, San Francisco
PHILIP K. HOPKE, Clarkson University, Potsdam, NY
SAMUEL KACEW, University of Ottawa, Ontario
H. SCOTT MATTHEWS, Carnegie Mellon University, Pittsburgh, PA
THOMAS E. MCKONE, University of California, Berkeley
TERRY L. MEDLEY, E.I. du Pont de Nemours & Company, Wilmington, DE
JANA MILFORD, University of Colorado at Boulder, Boulder
RICHARD L. POIROT, Vermont Department of Environmental Conservation, Waterbury
MARK A. RATNER, Northwestern University, Evanston, IL
KATHRYN G. SESSIONS, Health and Environmental Funders Network, Bethesda, MD
JOYCE S. TSUJI, Exponent Environmental Group, Bellevue, WA

Senior Staff

JAMES J. REISA, Director
DAVID J. POLICANSKY, Scholar
RAYMOND A. WASSEL, Senior Program Officer for Environmental Studies
ELLEN K. MANTUS, Senior Program Officer for Risk Analysis
SUSAN N.J. MARTEL, Senior Program Officer for Toxicology
EILEEN N. ABT, Senior Program Officer
MIRSADA KARALIC-LONCAREVIC, Manager, Technical Information Center
RADIAH ROSE, Manager, Editorial Projects

---

[1]This study was planned, overseen, and supported by the Board on Environmental Studies and Toxicology.

OTHER REPORTS OF THE
BOARD ON ENVIRONMENTAL STUDIES AND TOXICOLOGY

Science for Environmental Protection: The Road Ahead (2012)
Exposure Science in the 21st Century: A Vision and A Strategy (2012)
A Research Strategy for Environmental, Health, and Safety Aspects of Engineered Nanomaterials (2012)
Macondo Well–Deepwater Horizon Blowout: Lessons for Improving Offshore Drilling Safety (2012)
Feasibility of Using Mycoherbicides for Controlling Illicit Drug Crops (2011)
Improving Health in the United States: The Role of Health Impact Assessment (2011)
A Risk-Characterization Framework for Decision-Making at the Food and Drug Administration (2011)
Review of the Environmental Protection Agency's Draft IRIS Assessment of Formaldehyde (2011)
Toxicity-Pathway-Based Risk Assessment: Preparing for Paradigm Change (2010)
The Use of Title 42 Authority at the U.S. Environmental Protection Agency (2010)
Review of the Environmental Protection Agency's Draft IRIS Assessment of Tetrachloroethylene (2010)
Hidden Costs of Energy: Unpriced Consequences of Energy Production and Use (2009)
Contaminated Water Supplies at Camp Lejeune—Assessing Potential Health Effects (2009)
Review of the Federal Strategy for Nanotechnology-Related Environmental, Health, and Safety Research (2009)
Science and Decisions: Advancing Risk Assessment (2009)
Phthalates and Cumulative Risk Assessment: The Tasks Ahead (2008)
Estimating Mortality Risk Reduction and Economic Benefits from Controlling Ozone Air Pollution (2008)
Respiratory Diseases Research at NIOSH (2008)
Evaluating Research Efficiency in the U.S. Environmental Protection Agency (2008)
Hydrology, Ecology, and Fishes of the Klamath River Basin (2008)
Applications of Toxicogenomic Technologies to Predictive Toxicology and Risk Assessment (2007)
Models in Environmental Regulatory Decision Making (2007)
Toxicity Testing in the Twenty-first Century: A Vision and a Strategy (2007)
Sediment Dredging at Superfund Megasites: Assessing the Effectiveness (2007)
Environmental Impacts of Wind-Energy Projects (2007)
Scientific Review of the Proposed Risk Assessment Bulletin from the Office of Management and Budget (2007)
Assessing the Human Health Risks of Trichloroethylene: Key Scientific Issues (2006)
New Source Review for Stationary Sources of Air Pollution (2006)
Human Biomonitoring for Environmental Chemicals (2006)

Health Risks from Dioxin and Related Compounds: Evaluation of the EPA Reassessment (2006)
Fluoride in Drinking Water: A Scientific Review of EPA's Standards (2006)
State and Federal Standards for Mobile-Source Emissions (2006)
Superfund and Mining Megasites—Lessons from the Coeur d'Alene River Basin (2005)
Health Implications of Perchlorate Ingestion (2005)
Air Quality Management in the United States (2004)
Endangered and Threatened Species of the Platte River (2004)
Atlantic Salmon in Maine (2004)
Endangered and Threatened Fishes in the Klamath River Basin (2004)
Cumulative Environmental Effects of Alaska North Slope Oil and Gas Development (2003)
Estimating the Public Health Benefits of Proposed Air Pollution Regulations (2002)
Biosolids Applied to Land: Advancing Standards and Practices (2002)
The Airliner Cabin Environment and Health of Passengers and Crew (2002)
Arsenic in Drinking Water: 2001 Update (2001)
Evaluating Vehicle Emissions Inspection and Maintenance Programs (2001)
Compensating for Wetland Losses Under the Clean Water Act (2001)
A Risk-Management Strategy for PCB-Contaminated Sediments (2001)
Acute Exposure Guideline Levels for Selected Airborne Chemicals (fourteen volumes, 2000-2013)
Toxicological Effects of Methylmercury (2000)
Strengthening Science at the U.S. Environmental Protection Agency (2000)
Scientific Frontiers in Developmental Toxicology and Risk Assessment (2000)
Ecological Indicators for the Nation (2000)
Waste Incineration and Public Health (2000)
Hormonally Active Agents in the Environment (1999)
Research Priorities for Airborne Particulate Matter (four volumes, 1998-2004)
The National Research Council's Committee on Toxicology: The First 50 Years (1997)
Carcinogens and Anticarcinogens in the Human Diet (1996)
Upstream: Salmon and Society in the Pacific Northwest (1996)
Science and the Endangered Species Act (1995)
Wetlands: Characteristics and Boundaries (1995)
Biologic Markers (five volumes, 1989-1995)
Science and Judgment in Risk Assessment (1994)
Pesticides in the Diets of Infants and Children (1993)
Dolphins and the Tuna Industry (1992)
Science and the National Parks (1992)
Human Exposure Assessment for Airborne Pollutants (1991)
Rethinking the Ozone Problem in Urban and Regional Air Pollution (1991)
Decline of the Sea Turtles (1990)

Copies of these reports may be ordered from the National Academies Press
(800) 624-6242 or (202) 334-3313
www.nap.edu

Preface

Under the US Endangered Species Act (ESA), the Fish and Wildlife Service (FWS) and the National Marine Fisheries Service (NMFS) are responsible for designating species as endangered or threatened (that is, listing species) and determining whether federal actions might jeopardize the continued existence of a listed species or adversely affect its critical habitat. Under the Federal Insecticide, Fungicide, and Rodenticide Act (FIFRA), the US Environmental Protection Agency (EPA) is responsible for registering pesticides and ensuring that pesticides do not cause unreasonable adverse effects on the environment, which includes listed species and their critical habitats. Over the years, EPA, FWS, and NMFS have struggled unsuccessfully to reach a consensus on approaches to assessing the risks to listed species. Consequently, EPA, FWS, NMFS, and the US Department of Agriculture (USDA) asked the National Research Council to examine scientific and technical issues related to determining risks to species that are listed under the ESA posed by pesticides that are registered under FIFRA.

In this report, the Committee on Ecological Risk Assessment under FIFRA and ESA first provides a common approach that EPA, FWS, and NMFS could use to conduct assessments. It then discusses models, data, and uncertainties associated with exposure analysis and addresses various issues associated with assessing the effects of pesticides on listed species, including evaluating sublethal, indirect, and cumulative effects; modeling population-level effects; considering the effects of chemical mixtures; and incorporating uncertainties into the effects analysis. The committee closes by discussing the risk-characterization process and the need to propagate uncertainty through all components of the assessment so that decision-makers are well informed regarding the risk estimates produced.

The present report has been reviewed in draft form by persons chosen for their diverse perspectives and technical expertise in accordance with procedures approved by the National Research Council Report Review Committee. The purpose of the independent review is to provide candid and critical comments that will assist the institution in making its published report as sound as possible and to ensure that the report meets institutional standards of objectivity, evidence, and responsiveness to the study charge. The review comments and draft manuscript remain confidential to protect the integrity of the deliberative process. We thank the following for their review of this report: Steven Bartell, Cardno ENTRIX; May Berenbaum, University of Illinois; Nancy Bryson, Holland & Hart, LLP; Francesca Dominici, Harvard School of Public Health; Scott Ferson, Applied Biomathematics; Robert Gilliom, National Water Quality Assessment Program, USGS; Tilghman Hall, Bayer CropScience; Jeffrey Jenkins, Oregon State University; Andreas Kortenkamp, Brunel University; Bernalyn McGaughey, Compliance Services International; Anke Mueller-Solger, California Delta Stewardship Council; Terrance Quinn, University of Alaska Fairbanks; Joseph Rodricks, ENVIRON; Kenneth Rose, Louisiana State University; and Janet Silbernagel, University of Wisconsin, Madison.

Although the reviewers listed above have provided many constructive comments and suggestions, they were not asked to endorse the conclusions or recommendations, nor did they see the final draft of the report before its release. The review of the report was overseen by the review coordinator, Danny Reible, The University of Texas at Austin, and the review monitor, Michael Ladisch, Purdue University. Appointed by the National Research Council, they were responsible for making certain that an independent examination of the report was carried out in accordance with institutional procedures and that all review comments were carefully considered. Responsibility for the final content of the report rests entirely with the committee and the institution.

One committee member, Daniel Goodman, disagreed with the committee on several points and prepared a dissenting statement that was included as an appendix in the draft report that was submitted to peer reviewers. The report has been substantially revised in response to reviewer comments, and many issues raised by Dr. Goodman have been addressed with changes to the report. However, Dr. Goodman passed away while the report was in review, so determining how he would have judged the revised report is not possible. Accordingly, his dissenting statement has not been included in this final report; however, it is available in the public access file.

The committee gratefully acknowledges the following for their presentations to the committee during open sessions: Ann Bartuska, David Epstein, and Harold Thistle, USDA; Steven Bradbury and Edward Odenkirchen, EPA; Aimee Code, Northwest Center for Alternatives to Pesticides; Nancy Golden, FWS; Christian Grue, University of Washington; Barbara Harper, Confederated Tribes of the Umatilla Indian Reservation; Scott Hecht and Nathaniel Scholz, National Oceanic and Atmospheric Administration; Jeffrey Jenkins, Oregon State University; Steve Mashuda, Earthjustice; Bernalyn McGaughey, Compliance Services International; John Stark, Washington State University; and Mike Willett, Northwest Horticultural Council. The committee members also thank the staff of EPA, FWS, and NMFS for being so helpful in answering their numerous questions throughout the study process. It especially thanks Jim Cowles, formerly of the Washington State Department of Agriculture, and Scott McMurry, Oklahoma State University, for their useful input in the early deliberations of this study.

Preface

The committee is grateful for the assistance of the National Research Council staff in preparing this report. Staff members who contributed to the effort are Ellen Mantus and David Policansky, project codirectors; Keri Stoever, research associate; James Reisa, director of the Board on Environmental Studies and Toxicology; Norman Grossblatt, senior editor; Mirsada Karalic-Loncarevic, manager of the Technical Information Center; Radiah Rose, manager of editorial projects; and Craig Philip, senior program assistant.

I especially thank the members of the committee for their efforts throughout the development of this report.

> Judith E. McDowell, *Chair*
> Committee on Ecological Risk
> Assessment Under FIFRA and ESA

Dedication

This report is dedicated to Dr. Daniel Goodman (1945-2012), who served on the committee that authored this report until November 14, 2012, when he passed away unexpectedly. Dr. Goodman was professor and director of the Environmental Statistics Group in the Department of Ecology at Montana State University in Bozeman, where he had been on the faculty since 1980. Dr. Goodman provided advice to several federal agencies, including NOAA and EPA, and had served as a report reviewer for the NRC before becoming a member of this committee. The committee and the NRC are grateful for his service and his contributions.

Contents

SUMMARY 3

1 INTRODUCTION 16
   The Federal Insecticide, Fungicide, and Rodenticide Act, 16
   The Endangered Species Act, 20
   The Relationship Between The Two Acts, 23
   The Committee and Its Task, 25
   The Committee's Approach to Its Task, 25
   Organization of the Report, 27
   References, 27

2 A COMMON APPROACH AND OTHER OVERARCHING
   ISSUES 28
   A Common Approach, 28
   Coordination among Agencies, 34
   Uncertainty, 37
   Best Data Available, 39
   Conclusions and Recommendations, 43
   References, 45

3 EXPOSURE ANALYSIS 49
   Exposure-Modeling Practices, 49
   Geospatial Data for Habitat Delineation and Exposure Modeling, 55
   Uncertainties in Exposure Modeling and Parameter Inputs, 65
   Conclusions and Recommendations, 79
   References, 81

4 EFFECTS ANALYSIS 91
   Sublethal, Indirect, and Cumulative Effects, 92
   Effects Models, 101
   Mixtures, 108
   Interspecies Extrapolations and Surrogate Species, 128
   Other Uncertainties in Effects Analysis, 131

Conclusions and Recommendations, 132
References, 135

5   **RISK CHARACTERIZATION** 148
Concentration-Ratio Approach, 149
Probabilistic Approach, 150
Conclusions, 152
References, 152

APPENDIXES

A   **SELECTED EXCERPTS FROM 40 CFR PART 158 – DATA REQURIMENTS FOR PESTICIDES**... 155

B   **BIOGRAPHICAL INFORMATION ON THE COMMITTEE ON ECOLOGICAL RISK ASSESSMENT UNDER FIFRA AND ESA** 171

BOXES, FIGURES, AND TABLES

BOXES

1-1   Statement of Task, 26
2-1   Generic Outline for Reporting Ecological Risk-Assessment Results for Listed Species or Their Critical Habitats, 36
3-1   AgDRIFT Inputs, 71
4-1   Ecological Risk Assessment in Species That Have Complex Population Structure and Life History: Pacific Salmon and Trout, 96

FIGURES

S-1   Relationship between the Endangered Species Act (ESA) process and the ecological risk assessment (ERA) process, 6
1-1   Consultation process under ESA Section 7 for a federal action that potentially could affect a listed species or critical habitat, 22
2-1   Relationship between the Endangered Species Act (ESA) Section 7 decision process and the ecological risk assessment (ERA) process for a chemical stressor, 30
3-1   Organic-carbon normalized sorption coefficients, K_{oc}, values for atrazine plotted on a logarithmic scale, 72
3-2   Distribution of K_{oc} values for atrazine in a 6.25 hectare field, showing a range of about a factor of 2, 73

3-3 Upper panels: Distribution of observed (pseudo-) first-order biodegradation rates (per day) of flumetsulam as reported for 21 test soils by Lehman et al. (1992) on linear (left) and logarithmic (right) scales. Lower panels: Distribution of observed bacterial-number-normalized biodegradation rates (L/organism-hour) of the butoxyethyl ester of 2,4-D as reported for 33 test surface waters by Paris et al. (1981) on linear (left) and logarithmic (right) scales, 76
4-1 The effect of pesticide exposure on a density-dependence function, 109
4-2 Concentration-response curve of a chemical in the presence and absence of a synergist, 115
4-3 Example derivations used to determine K values, 127
4-4 Species sensitivity distributions for 2,2'-dipyridyldisulfide derived by using a Bayesian statistical model, 131

TABLES

S-1 Examples of Authoritative Sources of Geospatial Data, 11
2-1 Steps in the ESA Process as Related to Elements in the ERA Process for Pesticides, 31
3-1 Nested Hierarchy of Hydrologic Units, 60
3-2 Variability of Pesticide Degradation Rates in Soils, 74
3-3 Biodegradation Rate Coefficients and Other Physical-Chemical Data Used in PRZM/EXAMS Fate Modeling of the Ethylhexyl Ester of 2,4 D, 78
4-1 Example Dataset Used to Assess Exposure to and Effects of Cypermethrin, in Mixture with Several Other Chemicals, on the Sockeye Salmon, *Oncorhynchus nerka*, 126

ASSESSING RISKS TO
Endangered and Threatened Species
FROM
PESTICIDES

Summary

Under the Endangered Species Act (ESA), the US Fish and Wildlife Service (FWS) and the National Marine Fisheries Service (NMFS)—herein called the Services—are responsible for listing species as endangered or threatened and for designating critical habitats that are essential for their conservation. Furthermore, in consultation with the Services, federal agencies must ensure that their actions are not likely to jeopardize listed species or adversely affect critical habitats. Under the Federal Insecticide, Fungicide, and Rodenticide Act (FIFRA), the US Environmental Protection Agency (EPA) is responsible for registering or reregistering pesticides and must ensure that pesticide use does not cause any unreasonable adverse effects on the environment, which is interpreted to include listed species and their critical habitats. The agencies have developed their own approaches to evaluating environmental risk, and their approaches differ because their responsibilities, institutional cultures, and expertise differ. Over the years, the agencies have tried to resolve their differences but have been unsuccessful in reaching a consensus regarding their assessment approaches. As a result, FWS, NMFS, EPA, and the US Department of Agriculture asked the National Research Council (NRC) to examine scientific and technical issues related to determining risks posed to listed species by pesticides. Specifically, the NRC was asked to evaluate methods for identifying the best scientific data available; to evaluate approaches for developing modeling assumptions; to identify authoritative geospatial information that might be used in risk assessments; to review approaches for characterizing sublethal, indirect, and cumulative effects; to assess the scientific information available for estimating effects of mixtures and inert ingredients; and to consider the use of uncertainty factors to account for gaps in data. The present report, which was prepared by the NRC Committee on Ecological Risk Assessment under FIFRA and ESA, is the response to that request.

THE FEDERAL INSECTICIDE, FUNGICIDE, AND RODENTICIDE ACT AND THE ENDANGERED SPECIES ACT

FIFRA is the federal statute that governs the sale, distribution, and use of pesticides in the United States; it assigns EPA the authority to issue pesticide registrations or reregistrations, which are required for use of the pesticides. To obtain a registration, an applicant must demonstrate that a pesticide will perform its intended function and will not cause unreasonable adverse environmental effects. Once granted, the registration requires that the pesticide be labeled with specific product information, directions for use, and hazard information; the label specifies legal use of the pesticide.

The ESA is the federal statute that assigns FWS and NMFS the authority to designate species as threatened or endangered—that is, to "list" species—and governs the activities that might affect listed species. Under the ESA, federal agencies must ensure that their actions do not harm listed species or jeopardize their existence. Accordingly, if EPA is deciding whether to register a pesticide, it must determine whether the action "may affect" a listed species. If the answer is yes, EPA has the option of initiating a formal consultation or conducting further analysis to determine whether the action is "likely to adversely affect" listed species. If EPA determines that the action is not likely to affect listed species adversely—and FWS or NMFS, as appropriate, agrees—no further consultation is required. However, if EPA determines that the action is likely to affect a listed species adversely, a formal consultation is required, and FWS or NMFS must determine whether the proposed action is likely to jeopardize the existence of the listed species. The product of that determination is called a biological opinion (BiOp) and is issued by FWS or NMFS.

Compliance with the ESA in registering pesticides creates some challenges. First, pesticides are intended to harm target organisms and are intentionally released into the environment. Other species that are in an area where a pesticide is applied could be exposed to and harmed by the pesticide. Second, FIFRA requires that EPA must determine before registering a pesticide that the use of the pesticide will not cause an unreasonable adverse effect on the environment, taking into account economic and social benefits associated with its use. That is, EPA weighs the costs to human health and the environment that could result from pesticide use against social and economic benefits, such as the benefits of mitigating disease vectors and reducing crop damage. The ESA prohibits jeopardizing listed species or adversely affecting their critical habitats but does not generally consider economic and social costs and benefits. Third, FIFRA creates a national registration process in which pesticides are registered on a nationwide basis, but the ESA calls for evaluating effects on specific species and their critical habitats and thus is geographically and temporally focused. The differences between the statutes have led to conflicting approaches in evaluating risks and have contributed to the current inability to reach consensus on assessing risks to listed species from pesticides.

A COMMON APPROACH

Compliance with the ESA in the context of pesticide registration requires EPA and the Services to determine the probability of adverse effects on listed species and their critical habitat when a pesticide is used according to its label requirements. Clearly, there are tensions among the agencies in making that determination, many of which seem to result from different assumptions, technical approaches (data and models used), and risk-calculation methods. What is needed is a common, scientifically credible approach that is acceptable to EPA and the Services. The committee concludes that the risk-assessment paradigm that traces its origins to the seminal NRC report *Risk Assessment in the Federal Government: Managing the Process*[1] and more recently to the NRC report *Science and Decisions: Advancing Risk Assessment*[2] provides such an approach. After 30 years of use and refinement, this risk-assessment paradigm has become scientifically credible, transparent, and consistent; can be reliably anticipated by all parties involved in decisions regarding pesticide use; and clearly articulates where scientific judgment is required and the bounds within which such judgment can be applied. The process is used for human-health and ecological risk assessments and is used broadly throughout the federal government. Thus, the committee concludes that the risk-assessment paradigm reflected in the ecological risk assessment (ERA) process is singularly appropriate for evaluating risks posed to ecological receptors, such as listed species, by chemical stressors, such as pesticides.

Figure S-1 shows the three major steps in the ESA process in connection with the ERA framework. As illustrated in the figure, the framework is the same at each step, but the contents of each element (problem formulation, exposure and effects analysis, and risk characterization) are expected to change as the focus shifts from assessing whether a pesticide "may affect" a listed species (Step 1) to whether it is "likely to adversely affect" a listed species (Step 2) to whether it is likely to jeopardize the continued existence of a listed species (Step 3). That is, the assessment becomes more focused and specific to the chemicals, species, and habitats of concern as it moves from Step 1 to Step 3. If the Services can build on the EPA assessment conducted for Steps 1 and 2 rather than conducting a completely new analysis for Step 3, the ERA will likely be more effective and scientifically credible. Although the committee does not expect the basic risk-assessment framework to change, it recognizes that risk-assessment approaches and methods for determining, for example, what is hazardous, what concentration or quantity is hazardous, what end points constitute an adverse effect, and when, where, and how much exposure is occurring will continue to evolve.

Given the changing scope of the ERA process from Step 1 to Step 3, EPA and the Services need to coordinate to ensure that their own technical needs are met. One approach is to use problem formulation, conducted as part of the ERA process, as an effective means for agencies to coordinate and reach agreement

---

[1] NRC (National Research Council). 1983. Risk Assessment in the Federal Government: Managing the Process. Washington, DC: National Academy Press.

[2] NRC (National Research Council). 2009. Science and Decisions: Advancing Risk Assessment. Washington, DC: National Academies Press.

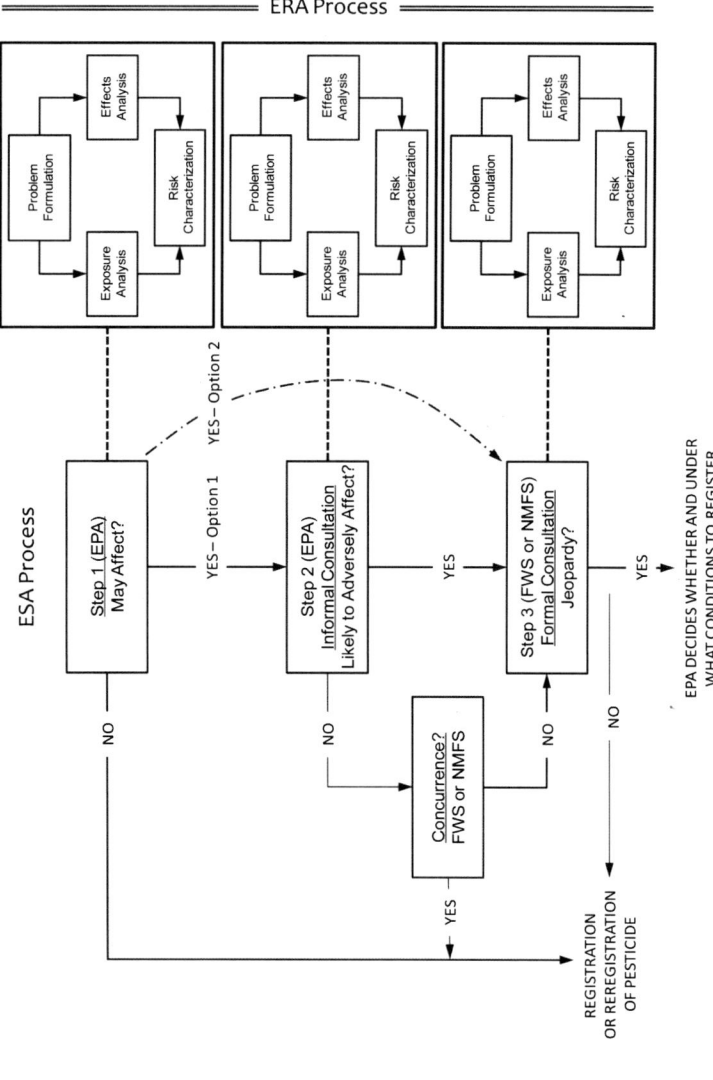

FIGURE S-1 Relationship between the Endangered Species Act (ESA) process and the ecological risk assessment (ERA) process. Each step answers the question that appears in the box.

on many of the key technical issues involved in assessing risks posed to listed species by pesticides. Another approach would be to use technical working groups that address technical details of the assessment approach and other working groups to address policy-based issues. Regardless of the approach, the committee views coordination among EPA and the Services as a collegial exchange of technical and scientific information for the purpose of producing a complete and representative assessment of risk that includes a discussion on the types and depths of analyses needed for the decision and on the time and resources available.

BEST DATA AVAILABLE

One of the critical tasks in any risk assessment is to identify the data that will be used. The ESA directs the Services to conduct assessments on the basis of the "best scientific and commercial data available." However, the ESA, its legislative history, the rules and policies of the Services, and court cases contain little guidance for elaborating the meaning of that mandate, and the agencies do not appear to have formal protocols that define "best data available." Consequently, there have been some conflicts about what data to include in the assessments. EPA and the Services do have information-quality guidelines, and each appears to use assessment factors that include data-quality and data-relevance criteria.

Regardless of the breadth of the data collection, some guidelines—such as those listed below—need to be followed in identifying and selecting data for a credible assessment.

- *Document the strategy for all data searches and retrieval.* For example, if a repository database is searched, the date that the search was conducted and all search terms and search criteria should be documented. The content and scope of the repository, its criteria for inclusion and exclusion of data, the periodicity of its updates, and its quality-control measures also should be documented.
- *To ensure that the best data available are used, screen the data first for relevance.*[3] Information that is not relevant clearly should not be used in assessing risk. Data should be from studies of the species and chemicals being assessed, or there should be a reasonable theoretical basis for data extrapolation. The data should also be applicable to the locations being considered and should be recent.
- *Review the quality of the relevant data before they are used in a risk assessment.*[4] Sufficient information should be included to enable an independent

---

[3]Relevance refers to the applicability of the data for the intended purpose.

[4]Quality refers to the scientific adequacy of the design and execution of data collection, the analyses that use the data, and the data reporting.

evaluation of data quality. Data sources that lack sufficient details for adequate scientific evaluation—such as poster presentations, abstracts, anecdotal or personal communications, and secondary sources—might provide useful background knowledge or support an overall weight-of-evidence evaluation but should not be the sole basis of conclusions in an assessment.

- *For transparency, document the evaluation of all data used with particular attention to sources, relevance, and quality and describe any issues associated with those data attributes in the discussion of uncertainty in the risk characterization.*

Given that various stakeholders are aware of and can provide relevant and high-quality data, the committee encourages provision for their involvement in the early stages and throughout the risk-assessment process. The committee notes that stakeholder data are expected to meet the same standards of relevance and quality as all other data.

EXPOSURE

Exposure analysis is a principal component of ERA and involves estimating the concentrations of various chemicals released into the environment and their breakdown products of toxicological significance. The following sections discuss exposure-modeling practices and the criteria for authoritative geospatial data and highlight the committee's conclusions on those topics.

Exposure-Modeling Practices

To determine whether a pesticide will adversely affect or jeopardize the existence of a listed species or its critical habitat, one must estimate the concentration to which the species might be exposed or the concentration that might result in the ecosystem. To accomplish that task, chemical fate and transport models are used. Because the pathways by which pesticides move from their points of application to habitats of listed species can involve a complex sequence of transfers with diverse degradation processes, it is common to use a linked series of models to estimate exposure.

The committee acknowledges that the models used for exposure analyses have several strengths but emphasizes that a model's limitations need to be recognized and the model used in the appropriate context. As noted above, the committee has suggested a common approach that involves more refined and sophisticated modeling and analysis as one moves from Step 1 to Step 3 in the ESA process. Given the current practices in exposure analysis and the need to estimate pesticide exposures and the associated spatial-temporal variations experienced by listed species and their habitats, the committee envisions the following stepwise approach to exposure modeling.

- *Step 1 (EPA)*. Initial exposure modeling would answer the question, Do the areas where the pesticide will be used overlap spatially with the habitats of any listed species? The Services, which have extensive knowledge of the natural history of listed species, could help EPA to identify overlaps of areas where a pesticide might be used and the habitats of listed species.
- *Step 2 (EPA)*. If area overlaps are identified in Step 1, EPA would confer with the Services to identify relevant environmental compartments (water, soil, air, and biota), associated characteristics, and critical times or seasons in which environmental exposure concentrations need to be estimated. If the models indicate that substantial amounts of pesticides move off the application site and into the surrounding ecosystems, more sophisticated fate and transport processes could be used. At that point, the fate model could be simplified to remove processes that are unimportant in the specific regions where the listed species are and set up to estimate time-varying and space-varying pesticide concentrations in typical habitats with associated uncertainties. On the basis of the modeling results, EPA could then make a decision about the need for formal consultation with the Services.
- *Step 3 (Services)*. During a formal consultation, the Services would further refine the exposure models to develop quantitative estimates of pesticide concentrations and their associated distributions for the particular listed species and their habitats. To that end, the models would use site-specific input values, such as actual pesticide application rates, locally relevant geospatial data, and time-sensitive life stages of listed species.

The committee emphasizes that many parameters are used in chemical fate and transport models, and their accuracy is important ultimately to the concentrations estimated in the modeling efforts. Little effort has been expended in evaluating the data inputs relevant to particular ESA evaluations. Therefore, if the agencies want to obtain more accurate modeling results, a subset of case-specific exposure estimates should be evaluated by pursuing a measurement campaign specifically coordinated with several pesticide field applications. The committee notes that field studies need to be distinguished from general monitoring studies. General monitoring studies provide information on pesticide concentration on the basis of monitoring of specific locations at specific times and are not associated with specific applications of pesticides under well-described conditions. Therefore, general monitoring data cannot be used to estimate pesticide concentrations after a pesticide application or to evaluate the performance of fate and transport models.

Geospatial Data, Habitat Delineation, and Exposure Analysis

Habitat includes all environmental attributes present in an area that allow an organism to survive and reproduce, and habitat delineation is necessary for determining where a pesticide and a species might co-occur, for calculating spatially explicit estimates of pesticide exposure, and for specifying the spatial structure of population models used in effects analysis (see below). Several methods for identifying and statistically modeling associations between species and their environment exist; although some caveats and uncertainties are associated with them, quantitative statistical habitat delineation is typically objective and more reliable than qualitative and subjective habitat descriptions.

The accuracy and reliability of habitat delineation and exposure analysis are increased substantially by the use of authoritative geospatial data. To be considered authoritative, geospatial data on any scale need to meet three criteria: availability from a widely recognized and respected source, public availability, and inclusion of metadata[5] that are consistent with the standards of the National Spatial Data Infrastructure—a federal interagency program to organize and share spatial data and to ensure their accuracy. The geospatial data that are most useful for delineating habitat and estimating exposure are data on topography, hydrography, meteorology, solar radiation, soils, geology, and land cover. Table S-1 provides some examples of authoritative sources of those data. In many cases, there are multiple authoritative sources for each type of data on different spatial and temporal scales. Although it would be ideal to be able to identify specific authoritative sources, no one source will be best for all habitat delineations, exposure analyses, or other applications. However, accuracy assessments that generally are available for authoritative data sources might allow one to gauge which source is likely to be the most reliable for a particular objective.

EFFECTS

Pesticides are designed to have biological activity; specifically, they are "intended for preventing, destroying, repelling or mitigating" pests. Consequently, they have the potential to cause a variety of effects on nontarget organisms, including listed species. Determining the potential for and possible magnitude of effects is a process known as effects analysis. The following sections consider various topics on effects analysis as they are related to the committee's task and highlight the committee's conclusions on the topics.

Sublethal, Indirect, and Cumulative Effects

Pesticides can kill organisms but can also affect reproduction or growth or make organisms less competitive. Although EPA and the Services agree that those sublethal (less-than-lethal) effects should be considered in the assessment process, they disagree on the extent to which they can be included. To address

---

[5]Metadata document the fundamental attributes of data, such as who collected them, when and where they were collected, what variables were measured, how and in what units measurements were taken, and the coordinate system used to identify locations.

that issue, the committee first considered how to define objectively the degree to which observed effects are adverse. Defining adversity is essential for ERA because the mere existence of an effect is not sufficient to conclude that it is adverse. The committee concluded that the only way to determine whether an effect is adverse and how adverse it might be is to assess the degree to which it affects an organism's survival and reproductive success; any effect that results in a change in either survival or reproduction is relevant to the assessment, and any effect that does not change either outcome is irrelevant with respect to a quantitative assessment of population effects. Thus, EPA in Step 2 (see Figure S-1) should conduct a broad search to identify sublethal effects of pesticides and any information on concentration-response relationships. In Step 3, the Services should then show how such effects change probability of survival or reproduction of the listed species and incorporate such information into the population viability analyses or state that such relationships are unknown but possible and include a qualitative discussion in the uncertainty section of the BiOp. The inability to quantify the relationship between a sublethal effect and survival or reproductive success does not mean that the sublethal effect has no influence on population persistence; but in the absence of data, the relationship remains a hypothesis that can be discussed only qualitatively with reference to the scientific literature to explain why such a hypothesis is tenable.

TABLE S-1 Examples of Authoritative Sources of Geospatial Data

Data Type	Examples of Authoritative Data Sources
Topography	Topographic features can be derived from elevation data in the National Elevation Dataset, the Shuttle Radar Topography Mission, and the Global Digital Elevation Map.
Hydrography	Watershed data are available on line from EPA; watersheds are referred to by hydrologic unit codes of the US Department of Agriculture Natural Resources Conservation Service.
Meteorology	Data are available from the National Oceanographic and Atmospheric Administration National Climatic Data Center.
Solar radiation	Solar-radiation data are available from the National Aeronautics and Space Administration Earth Observing System Solar Radiation and Climate Experiment;[a] solar insolation can be estimated by using the on-line calculator of the Photovoltaic Education Network.
Soils	Soil surveys are available from the US Department of Agriculture Natural Resources Conservation Service.
Geology	Geological data are available from the US Geological Survey Mineral Resources Online Spatial Data.
Land cover	Land-cover data are available from the National Agricultural Statistics Service.

[a]Solar-radiation measurements are taken at the top of Earth's atmosphere. Computer modeling is required to estimate solar radiation at Earth's surface.

In most cases, pesticides have the potential to affect a listed species indirectly—not through direct exposure but through effects on other species in the community. For example, the prey of a listed species might be reduced in abundance or eliminated by the pesticide, and this would affect the survival of the species. As in the case of sublethal effects, EPA and the Services differ about the degree to which indirect effects can be included in an assessment. The committee recommends that indirect effects that can be quantified relatively easily be incorporated into the effects analysis. However, determining and quantifying most indirect effects can be challenging and can require complex models. When such modeling is conducted, uncertainties should be estimated quantitatively in a realistic and scientifically defensible manner and should be propagated formally and explicitly through the analysis.

A risk assessor must also consider cumulative effects. They are defined by regulation under the ESA as "those effects of future State or private activities, not involving Federal activities that are reasonably certain to occur within the action area of the Federal action subject to consultation" (50 CFR 402.02). However, cumulative effects typically are more broadly defined as effects that interact or accumulate over time and space. The committee could not determine a scientific basis for excluding past and present conditions (the environmental baseline) from the consideration of cumulative effects and therefore used that broad definition in its evaluation. The committee concluded that population models provide a framework for incorporating baseline conditions and projected future cumulative effects into an effects analysis.

One problem that arises in an effects analysis is how to extrapolate toxicity information on tested species to listed species. Although the idea of identifying an appropriate surrogate species is appealing, the committee finds such identification problematic because different species often respond differently to chemical exposures, and the sensitivity differences can be large. Furthermore, different life histories can complicate the extrapolation. A scientifically defensible alternative approach is to define a range of sensitivities within which the sensitivity of a listed species could reasonably be expected to occur or a range of sensitivities that could be used to make reasoned extrapolations from information on species that have been tested by using inferences based on other chemicals. Further details are provided in Chapter 4 of this report.

Effects Models

EPA and the Services use different approaches to determine the potential effects of a pesticide on a listed species and its critical habitat. EPA addresses population effects simply as extensions of individual effects: if survival or reproduction is affected, EPA assumes population-level consequences and enters consultation with the Services. The Services use population models to address the question of population persistence explicitly. Population models are used to estimate population-level end points—such as population growth rate, probabil-

Summary

ity of population survival (population viability), and probability of population recovery—on the basis of individual-level effects. For purposes of population modeling, the effects must be estimated at a range of concentrations that includes all values that the populations being assessed might plausibly experience. Therefore, test results expressed only as threshold values, such as a no-observed-adverse-effect level or a lowest observed-adverse-effect level, are insufficient for a population-level risk assessment.

Because the ESA is concerned with species, population models are necessary for quantifying the effects of pesticides on populations of listed species. Population models require three basic inputs: changes in survival or reproduction as a function of pesticide concentration, exposure estimates of pesticide concentration over time and space, and demographic and life-history information. There are a variety of population models, and the choice of a model will depend on the data available. Although species-specific models that incorporate all three inputs are preferred, in the absence of detailed demographic information it is reasonable to use simple generic models that characterize the life history of a group of species to estimate the effects of a pesticide on a given species. It is important to incorporate density dependence by using models with parameter values that are functions of population density or population size, but it is not accurate to assume that mortality due to pesticide exposure will be compensated for by density dependence because it is likely that such exposure will decrease the growth rate of the population at all densities and generally depress the curve of population size vs growth rate.

MIXTURES: AN IMPORTANT CONCERN FOR EXPOSURE AND EFFECTS ANALYSIS

Assessing the risks posed by exposure to mixtures is clearly a subject of disagreement and concern for the agencies. To address the mixture issue, the committee made several distinctions. First, some pesticides might contain more than one active ingredient (a chemical that is responsible for the biological effect of the pesticide); most pesticides contain other chemicals that are typically designated as "inerts."[6] Second, pesticides are often mixed with other chemicals before their application. The resulting mixtures are referred to as tank mixture and can contain other pesticides, fertilizers, and adjuvants—materials that facilitate handling and application, such as surfactants, compatibility agents, antifoaming agents, and drift-control agents. Third, chemicals from other sources are already in the environment; unless exposure occurs only at or near the point of pesticide application, species are more likely to be exposed to environmental mixtures than to a single pesticide formulation or tank mixture. Environmental

---

[6]The term *inerts* is defined by FIFRA as an ingredient that is not active. Inerts are intentionally added to pesticide products, and the term does not mean that the chemicals are nontoxic.

mixtures are formed when a tank mixture—active ingredients, inerts, and adjuvants—combines with other chemicals in the environment from other sources. Ideally, assessments should be based on exposure to all pesticide components and to other chemicals that are present in the exposure environment. However, quantitative estimates of exposure to environmental mixtures are difficult given the dynamic state of environmental mixtures over space and time. In any given location, the amounts of pesticide active ingredients, inerts, adjuvants, and other environmental chemicals are highly variable and depend on pesticide uses and other sources of environmental contamination.

EPA recognizes the potential importance of exposure to mixtures but typically assesses only pesticide active ingredients. The Services have expressed substantial concern about the need to account for mixture exposure but have dealt with the issue only with a qualitative discussion in their assessments. The greatest concern is that a mixture component might act to enhance the toxicity of a pesticide active ingredient. The committee notes that a quantitative assessment of the risk posed by chemical mixtures requires extensive data, including data on the identity, concentration, and toxicity of mixture components. Challenges in assessing risk to listed species posed by pesticide-containing mixtures arise largely because of the lack of such data *and* the lack of understanding of the potential for interactions among mixture components. In the absence of such quantitative data, the possible contribution of specific mixture components to the toxicity of a pesticide active ingredient cannot be incorporated into a quantitative risk assessment. The committee, however, emphasizes that the complexity of assessing the risk posed by chemical mixtures should not paralyze the process, and it provides guidelines in Chapter 4 of its report to help in determining when and how to consider components other than a pesticide active ingredient in a risk assessment.

RISK CHARACTERIZATION AND UNCERTAINTY

Risk characterization is the final stage of a risk assessment in which the results of the exposure and effects analyses are integrated to provide decision-makers with a risk estimate and its associated uncertainty. Two general approaches have been used for risk characterization: the risk-quotient (RQ) approach, which compares point estimates of exposure and effect values, and the probabilistic approach, which evaluates the probability that exposure to a chemical will lead to a specified adverse effect at some future time.

The RQ approach does not estimate risk—the probability of an adverse effect—itself but rather relies on there being a large margin between a point estimate that is derived to maximize a pesticide's environmental concentration and a point estimate that is derived to minimize the concentration at which a specified adverse effect is not expected. If the results raise doubts regarding the possibility of an adverse effect, the common response is to widen the margin by, for example, adding uncertainty factors or assuming more stringent, and possibly implausible, exposure scenarios. The flaw in that approach is that there is no accounting for what the probability of an adverse effect was before the application of assumptions, and there is no calculation of how their use modifies that probability. Accordingly, the committee concludes that adding uncertainty factors to RQs to account for lack of data (on formulation toxicity, synergy, additivity, or any other aspect) is unwarranted because there is no way to determine whether the assumptions that are used overestimate or underestimate the probability of adverse effects. Furthermore, the committee concludes that RQs are not scientifically defensible for assessing the risks to listed species posed by pesticides or indeed for any application in which the desire is to base a decision on the probabilities of various possible outcomes.

Instead, the committee recommends using a probabilistic approach that requires integration of the uncertainties (from sampling, natural variability, lack of knowledge, and measurement and model error) into the exposure and effects analyses by using probability distributions rather than single point estimates for uncertain quantities. The distributions are integrated mathematically to calculate the risk as a probability and the associated uncertainty in that estimate. Ultimately, decision-makers are provided with a risk estimate that reflects the probability of exposure to a range of pesticide concentrations and the magnitude of an adverse effect (if any) resulting from such exposure.

The committee recognizes the pragmatic demands of the pesticide-registration process and encourages EPA and the Services to consider the probabilistic methods that have already been successfully applied to pesticide risk assessments, that have otherwise appeared often in the technical literature, that are familiar to many risk-assessment practitioners, that can be implemented with commercially available software, and that are most readily explicable to decision-makers, stakeholders, and the public. The committee also recognizes that administrative and other nonscientific hurdles will need to be overcome to implement this approach, but moving the uncertainty analysis from the typical narrative addendum to an integral part of the assessment is possible and necessary to provide realistic, objective estimates of risk.

1

Introduction

The US Endangered Species Act (ESA) requires federal agencies to consult with the Fish and Wildlife Service (FWS) or the National Marine Fisheries Service (NMFS) when a federal action might affect a species that is listed as threatened or endangered (that is, a listed species) or its designated critical habitat. One such action that could potentially affect listed species or their critical habitats is the registration (or reregistration) of pesticides by the US Environmental Protection Agency (EPA) under the Federal Insecticide, Fungicide, and Rodenticide Act (FIFRA). Accordingly, EPA must first determine whether the registration (or reregistration) of a pesticide "may affect" a listed species. If so, EPA must initiate formal consultation *or* determine whether it is "likely to adversely affect" a listed species. If EPA determines that the pesticide registration is "not likely to adversely affect" a listed species—and FWS or NMFS, as appropriate, agrees—no further consultation is required. However, if EPA determines that the pesticide registration is "likely to adversely affect" a listed species, a formal consultation is required, and the product of that formal consultation is a biological opinion (BiOp) issued by FWS or NMFS. Over the last decade, several court cases have made it clear that formal or informal consultation is required when EPA registers or reregisters a pesticide that might affect a listed species. The consultations that have resulted from the court cases raise questions regarding the best approaches or methods for determining risks to listed species and their critical habitats. Because EPA, FWS, and NMFS have some fundamental differences in approaches, they and the US Department of Agriculture (USDA) asked the National Research Council (NRC) to examine scientific and technical issues related to determining risks to ESA-listed species from pesticides that are registered under FIFRA. As a result of the request, NRC convened the Committee on Ecological Risk Assessment under FIFRA and ESA, which prepared the present report.

THE FEDERAL INSECTICIDE, FUNGICIDE, AND RODENTICIDE ACT

FIFRA is the federal statute that governs the sale, distribution, and use of pesticides in the United States [7 U.S.C. §§ 136-136y]. EPA has the primary responsibility for administering FIFRA, and the states play an important role in enforcing the act. Under FIFRA, the term *pesticide* is defined as "any substance or mixture of substances intended for preventing, destroying, repelling or mitigating any pest" [7 U.S.C. § 136 (u)(1)].

Pursuant to FIFRA Section 3(a), a pesticide may not be sold or distributed in the United States without a license, known as a registration, from EPA. To obtain a FIFRA registration, an applicant must demonstrate, among other things, that the pesticide will "perform its intended function without unreasonable adverse effects on the environment" [§ 136a (c)(5)(C)] and that when the pesticide is "used in accordance with widespread and commonly recognized practice it will not generally cause unreasonable adverse effects on the environment" [§ 136a (c)(5)(D)]. FIFRA defines *environment* as "water, air, land, and all plants and man and other animals living therein and the interrelationships which exist among these" [§ 136 (j)]. It defines the phrase *unreasonable adverse effects on the environment* as any "unreasonable risk to man or the environment taking into account the economic, social, and environmental costs and benefits of the use of any pesticide" [§ 136 (z)(bb)(1)]. In other words, when deciding whether a particular pesticide meets the standard for registration, EPA must consider the economic and social benefits of using the pesticide and the risks to humans and the environment posed by its use. EPA has interpreted the "unreasonable adverse effects on the environment" standard to require a balancing of costs and benefits in which EPA weighs the costs to human health and the environment resulting from pesticide use against social and economic benefits, such as the benefits of mitigating disease vectors and reducing crop damage.

To obtain a registration, an applicant must provide data demonstrating that its pesticide does not cause unreasonable adverse effects. FIFRA does not mandate that any particular tests be conducted or that any particular type of data be submitted to obtain a registration. However, FIFRA Section 3(c)(2)(A) directs EPA to publish guidelines "specifying the kinds of information which will be required to support the registration of a pesticide" and directs EPA to revisit and revise these guidelines "from time to time." Pursuant to that section, EPA has promulgated rules in 40 C.F.R. Part 158 that establish data requirements for demonstrating that a particular pesticide product meets the standard for registration. Excerpts from Part 158 are provided in Appendix A of the present report. EPA has also developed a series of test guidelines that specify methods for conducting the studies that will generate the data to support registration.

Many of the data requirements in Part 158 address general information about a pesticide, such as its chemical composition and chemical and physical properties. Other data requirements focus on mammalian testing that can be used to evaluate the human health effects of pesticide exposure. Most important for purposes of this report, Part 158 includes a number of sections related to environmental risk, including risks to species that are not the targets of the pesticide (that is, nontarget species). For example, Subpart G requires avian oral toxicity testing, avian dietary toxicity testing, and avian reproduction testing and might require wild-mammal toxicity testing and simulated or actual field testing. Additional data on wildlife are required only case by case. Subpart G also requires acute toxicity tests on honeybees and various toxicity tests on freshwater fishes, freshwater invertebrates, and estuarine and marine organisms. Subpart L sets forth requirements for spray-drift data, and Subpart N sets forth requirements for environmental fate data, which are targeted at assessing "the presence of widely distributed and persistent pesticides in the environment which may result in loss of usable land, surface water, ground water, and wildlife resources, and…the potential environmental exposure of other nontarget organisms, such as fish and wildlife, to pesticides" [40 C.F.R § 158.130(h)(l)].

If, after evaluating the data submitted, EPA determines that the applicant has demonstrated that the standard for registration has been met, it will issue a registration. The registration will specify use restrictions that EPA has determined are necessary to meet the standard for registration. Most important, the registration will require that the pesticide be labeled with specific product information, directions for use, and hazard information. The product label dictates legal use of the pesticide. FIFRA provides that it is a violation of federal law "to use any registered pesticide in a manner inconsistent with its labeling" [§ 136 (j)(a)(2)(G)], and every registered pesticide product is required to bear a label containing this warning. Accordingly, the label is the vehicle not only for providing important information to end users but for mandating the purposes for which and the manner in which end users may use the pesticide product. The label instructions are necessary to ensure that the pesticide meets the standard for registration. A pesticide that might have an unreasonable adverse effect on the environment if used at a particular dosage, for a particular crop type, or in a particular manner might not have an unreasonable adverse effect if its use is restricted to other specified crops or specified application rates or restricted in other ways to minimize human health or environmental risks. Thus, the label language is EPA's primary regulatory tool for reducing pesticide risk under FIFRA. Users who fail to comply with label directions can incur penalties, although in practice it is extremely difficult to monitor every pesticide application to determine whether it was carried out according to the label.

Once a pesticide is registered, EPA does not require a permit or any other approvals before it is used. That is, there is no evaluation of specific pesticide applications; thus, the geographic and temporal factors specific to an application site or timing are not evaluated before the pesticide is released into the environment. However, some states have their own pesticide-permitting programs that apply to specific types of pesticide use (for example, aerial application). Furthermore, EPA has the authority under FIFRA to classify specific pesticides as "restricted use pesticides." Those pesticides can be used only under the supervision of a certified applicator who has received training in the proper handling and use of the pesticide in question. However, even when there are state permitting requirements and certified-applicator-training requirements, most pesticide use is regulated only by label restrictions without a requirement for a permit or other approval before use.

After a pesticide product is registered, FIFRA continues to impose responsibilities on the registrant, and EPA can require additional data submission. FIFRA Section 6(a)(2) requires that if at any time after the issuance of a registration a registrant obtains information that a pesticide has unreasonable adverse effects on the environment, the registrant is required to submit the information to EPA. And FIFRA Section 3(c)(2)(B) states that "if [EPA] determines that additional data are required to maintain in effect an existing registration of a pesticide, [EPA] shall notify all existing registrants of the pesticide to which the determination relates." If EPA invokes Section 3(c)(2)(B), referred to as a "data call-in," each registrant must provide evidence to EPA within 90 days that it is "taking steps to secure the additional data required." If EPA determines that a registrant has failed to take appropriate steps to secure the required data, it may initiate proceedings to suspend the registration of the pesticide. EPA can cancel a registration if it determines that a pesticide or its labeling does not comply with FIFRA or if the pesticide "generally causes unreasonable adverse effects on the environment when used in accordance with widespread and commonly recognized practice" (75 Fed. Reg. 68297[2010]). FIFRA Section 6(c) authorizes the suspension of a registration if EPA determines that suspension is necessary to prevent an imminent hazard during the time required for cancellation. FIFRA Section 2(l) defines *imminent hazard* to include a "situation which exists when the continued use of a pesticide during the time required for cancellation proceeding...will involve unreasonable hazard to the survival of a species declared endangered or threatened by the Secretary pursuant to the Endangered Species Act of 1973."

Congress has on several occasions directed EPA to review the human health and environmental effects of pesticides registered before some specified date. In 1972, revisions of FIFRA mandated that EPA re-evaluate registered pesticides—a process known as reregistration—by using current scientific and regulatory standards to ensure that the data used to register the pesticides originally meet current standards. In 1988, Congress imposed specific reregistration requirements that were intended to improve the speed and the nature of reregistration. The 1988 provisions established a multistep process with various deadlines intended to ensure that registrants submit required data to EPA in a timely manner. Under the 1988 amendments, failure to meet the data-submission deadlines could result in suspension or cancellation of a registration.

In 1996, Congress passed the Food Quality Protection Act (FQPA), which also amended FIFRA. The FQPA was focused on providing additional protections for humans, not wildlife, and required EPA to re-evaluate many food-use pesticides under new human-health standards. As a result of the re-evaluation, EPA canceled some pesticide uses, changed allowable application rates, and imposed use restrictions on others that were not aimed at reducing risk to wildlife but had that result.

THE ENDANGERED SPECIES ACT

The ESA is the federal statute that creates the authority to designate species as threatened or endangered and governs the activities that might affect those species (Endangered Species Act 16 U.S.C. §§ 1531–1544). The ESA is administered and enforced by two federal agencies that have jurisdiction for species in different ecosystems. FWS, in the Department of the Interior, typically is responsible for freshwater and terrestrial species, and NMFS, in the Department of Commerce, typically is responsible for marine and anadromous species (species that migrate from marine to freshwater environments to spawn, such as Pacific salmonids). The two agencies—referred to collectively as the Services and individually as the Service—are responsible for listing species as endangered or threatened under the ESA.

An *endangered species* is defined as a "species which is in danger of extinction throughout all or a significant portion of its range" [16 U.S.C. § 1532 (6)]. A *threatened species* is defined as a species that is "likely to become...endangered...within the foreseeable future throughout all or a significant portion of its range" [§ 1532 (20)]. Subspecies of "fish or wildlife or plants and any distinct population segment of any species of vertebrate fish or wildlife which interbreeds when mature" [§ 1532 (16)] are also included in the ESA's definition of species and thus can be listed. In this report, the terms *endangered species*, *threatened species*, and *listed species* can refer to subspecies or distinct population segments as defined by the ESA. Once a species is listed, the ESA requires that the Services designate critical habitat for each listed species. As of October 15, 2012, critical habitat had been designated for 653 of the 1,434 listed species that occur in the United States.

Endangered species are subject to several protections under the ESA, and threatened species are for the most part subject to the same protections. ESA Section 9 prohibits the "take" of listed species. The statute defines *take* as "harass, harm, pursue, hunt, shoot, wound, kill, trap, capture, or collect or attempt to engage in any such conduct" [16 U.S.C. § 1532 (19)]. The Services have further defined *harm* to include acts that involve substantial habitat modification or degradation that kills or injures listed species by substantially impairing essential behavior patterns, including breeding, feeding, and sheltering. That broad interpretation of harm has been upheld by the US Supreme Court [*Babbitt v Sweet Home Chapter of Communities for a Greater Oregon*, 515 U.S. 687, 698 (1995)]. The ESA authorizes the Services to assess penalties for unauthorized take of listed species and authorizes courts to impose injunctions to prevent a take from occurring or continuing. A federal agency (such as EPA) is liable for its actions, including, at least according to one court, the issuance of FIFRA registrations that result in a take of a listed species [*Defenders of Wildlife v Administrator, EPA*, 882 F. 2d 1294 (8th Cir. 1989)].

Section 7 of the ESA includes another important provision that specifically applies to actions of federal agencies. It mandates that federal agencies use their existing authorities to conserve endangered and threatened species and

Introduction

consult with the Services to ensure "that any action authorized, funded, or carried out by such agency is not likely to jeopardize the continued existence of any endangered species or threatened species or result in the destruction or adverse modification of [critical habitat] of such species" [16 U.S.C. § 1536 (a)(2)]. The phrase "jeopardize the continued existence of [a listed species]" means "to engage in an action that reasonably would be expected, directly or indirectly, to reduce appreciably the likelihood of both the survival and recovery of a listed species in the wild by reducing the reproduction, numbers, or distribution of that species" [50 CFR § 402.02].

Any proposed federal agency action that "may affect" listed species is subject to ESA Section 7 and could require a formal consultation (see Figure 1-1). The term "may affect" is defined broadly to include beneficial and adverse effects. For any action that "may affect" listed species, the action agency has two options: it may choose to initiate formal consultation or may determine whether the action is "likely to adversely affect" listed species. If the action agency determines, with written concurrence of FWS or NMFS, that the action is "not likely to adversely affect" a listed species or its critical habitat, no further consultation is required. However, if the action agency determines that the action is "likely to adversely affect" a listed species or its critical habitat, formal consultation is required. Through the formal consultation process, FWS or NMFS determines whether the proposed federal agency action is likely to jeopardize listed species; if so, FWS or NMFS will develop "reasonable and prudent alternatives" (RPAs) that, if implemented, are expected to avoid jeopardy. It is at the action agency's discretion whether to adopt the RPAs. However, the agency will be liable under Section 9 if a take results from its action and the take was not provided for by an incidental take statement (ITS) in the BiOp, the final document issued by FWS or NMFS. An ITS describes actions that will not be considered prohibited takes and describes "reasonable and prudent measures" that must be complied with to be covered by the ITS.

Unlike FIFRA and its implementing regulations, the ESA does not prescribe specific studies that must be conducted or specific data that must be collected or submitted in the consultation process. Instead, in several provisions of the ESA, Congress has directed the Services to make determinations based on the "best scientific and commercial data available." Similarly, the Services' rules on consultation state that

> the Federal agency requesting formal consultation shall provide the Service with the best scientific and commercial data available or which can be obtained during the consultation for an adequate review of the effects that an action may have upon listed species or critical habitat. This information may include the results of studies or surveys conducted by the Federal agency or the designated non-Federal representative. The Federal agency shall provide any applicant with the opportunity to submit information for consideration during the consultation [50 C.F.R. 402.14(d)].

In formulating its biological opinion, any reasonable and prudent alternatives, and any reasonable and prudent measures, the Service will use the best scientific and commercial data available [50 C.F.R. 402.14(g)(8)].

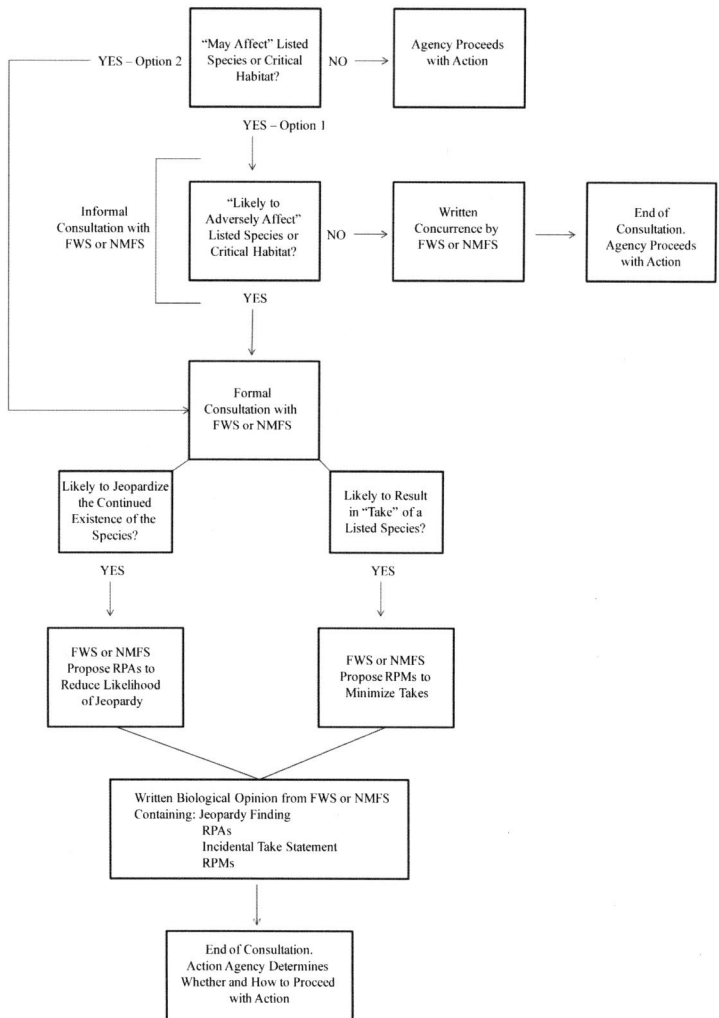

FIGURE 1-1 Consultation process under ESA Section 7 for a federal action that potentially could affect a listed species or critical habitat. If the agency determines that the action "may affect" the listed species or critical habitat, it has two options: (1) determine whether the action is "likely to adversely affect" or (2) go directly to formal consultation with the appropriate Service. Abbreviations: FWS, Fish and Wildlife Service; NMFS, National Marine Fisheries Service; RPA, Reasonable and Prudent Action; RPM, Reasonable and Prudent Measure.

The Services have also issued two policy statements on implementing the "best scientific and commercial data available" mandate. The first is the *Notice of Interagency Cooperative Policy on Information Standards* [59 Fed. Reg. 34271 (July 1, 1994)]. It applies to, among other things, decisions made in the Section 7 consultation process and states that biologists employed by the Services must evaluate all information to "ensure that any information used by the Services to implement the Act is reliable, credible, and represents the best scientific and commercial data available." It also expresses a preference that the Services use primary and original sources of information as the basis of its recommendations.

The second policy statement is the *Notice of Interagency Cooperative Policy for Peer Review in Endangered Species Act Activities* [59 Fed. Reg. 3270 (July 1, 1994)]. It provides that in making listing decisions and developing recovery plans under the ESA, the Services will seek independent peer review. It does not explicitly apply to decisions made in the Section 7 consultation process.

Neither the ESA nor its implementing regulations or policies provide detailed guidance on what is meant by "best scientific and commercial data available." Moreover, the legislative history of the ESA does not provide any clear direction on what Congress intended by using that language. However, experts who have studied the ESA, its legislative history, and circumstances surrounding the passage of the act have stated that the "best scientific and commercial data" mandate was generally intended to "ensure objective, value-neutral decision making by specially trained experts" (Doremus 2004). As one expert has opined, "taking the best available science mandate at face value, its most obvious purpose would seem to be to ensure that agency decisions are substantially as 'good' as can be" (Doremus 2004). Experts who have analyzed the case law involving the use of the best-available-science mandate have concluded that the cases suggest "no consistent thread or logic" (Brennan et al. 2003). Thus, there is little guidance in the ESA, its legislative history, the Services' rules and policies, or court cases to elaborate the meaning of the "best scientific and commercial data available" mandate in the ESA.

THE RELATIONSHIP BETWEEN THE TWO ACTS

At least one court has held that EPA can be liable for a take under the ESA if its registration of a pesticide results in the take of a listed species [*Defenders of Wildlife v Administrator, EPA*, 882 F. 2d 1294 (8th Cir. 1989)]. More important for the purposes of the present report, courts have held that EPA is required to comply with the ESA Section 7 consultation process when registering or taking other regulatory actions on pesticides under FIFRA. The requirement that EPA comply with the ESA when registering pesticides under FIFRA presents a number of challenges. First, pesticides, by their very nature, are intended to harm or disrupt a living organism in some way. Pesticides intended for outdoor agriculture, forestry, weed control, and other uses are also intentionally released into the environment. Consequently, if any listed species nest, roost, migrate through, or otherwise exist in a particular geographic location where pesticides are released, they could be exposed to potentially harmful substances, and takes could occur.

As described above, the ESA prohibits any take of a listed species and requires formal consultation for any agency action that is likely to affect any listed species adversely. FIFRA, in contrast, requires a cost-benefit balancing of the risks associated with the use of a pesticide and the social and economic benefits to be gained by its use. The ESA prohibits takes of listed species and seeks to ensure that federal agency actions do not jeopardize the continued existence of a listed species. Economic considerations do not come into play in ESA listing, take, or jeopardy evaluations as they do under FIFRA. The FIFRA cost-benefit standard applies whether or not listed species are at issue, although presumably harm to a listed species would be considered a high cost. In fact, the only place where FIFRA mentions threatened or endangered species is in Section 6(c)(1) of FIFRA, which authorizes EPA to "suspend the registration of a pesticide [if that] is necessary to prevent an imminent hazard during the time required for a cancellation proceeding." As noted above, FIFRA Section 2(l) defines *imminent hazard* to include a "situation which exists when the continued use of a pesticide during the time required for cancellation proceeding . . . will involve unreasonable hazard to the survival of a species declared endangered or threatened." FIFRA does not provide EPA with any other direction concerning listed species.

Another challenge for EPA in complying with the ESA for pesticide registrations is that FIFRA creates a national registration process whereas the ESA requires an evaluation of effects on the habitat of a listed species and individual members of a species. Under FIFRA, pesticide registration or cancellation decisions are made on a nationwide basis. The ESA, in contrast, is geographically and temporally focused. Although EPA typically considers geographic fate and exposure scenarios relevant to where and when a pesticide is expected to be used, it is challenging to design label restrictions and warnings to ensure that there is never an effect on a listed species.

Another difference between FIFRA and the ESA concerns data available for assessments. As indicated above, FIFRA requires the submission of data before registration, whereas under the ESA the Services are mandated to rely on the best data available (as opposed to requesting new data). Furthermore, under the ESA, decisions are not to be delayed because of a lack of data.

The differences between the statutes have led EPA and the Services to develop different approaches to ecological risk assessment that have often made it difficult for them to reach a scientific agreement. As a result, EPA and the Services decided to seek advice from the NRC on several scientific issues related to conducting an ecological risk assessment.

Introduction

THE COMMITTEE AND ITS TASK

The committee that was convened in response to the request from EPA, FWS, NMFS, and USDA included experts on salmonid biology, ecology, hydrology, geospatial analysis, exposure analysis, toxicology, population dynamics, statistics, uncertainty analysis, environmental law, and ecological, pesticide, and mixture risk assessment (see Appendix B for biographical information). The committee was asked to evaluate EPA's and the Services' methods for determining risks to listed species posed by pesticides and to answer questions concerning the identification of the best scientific data, the toxicological effects of pesticides and chemical mixtures, the approaches and assumptions used in various models, the analysis of uncertainty, and the use of geospatial data. See Box 1-1 for a verbatim statement of the committee's task.

THE COMMITTEE'S APPROACH TO ITS TASK

The committee held five meetings to assist it in accomplishing its task. The first three included open sessions during which the committee heard from the sponsors and invited speakers from academe, professional organizations, nonprofit organizations, and consulting agencies. The committee submitted written questions to the sponsors to clarify the charge questions, discussed their responses in an open session, and reviewed extensive literature on various aspects of ecological risk assessment and materials provided by the sponsors and stakeholders. As directed in its statement of task, the committee used the recent consultations between the NMFS and EPA as a reference for its evaluation of assessment methods used by EPA and the Services. It emphasizes that it did not specifically evaluate the biological opinions or EPA's effect determinations on Pacific salmonids; that would have been outside its charge. For ease of discussion, the committee has designated the steps in the ESA process—"may affect," "likely to adversely affect," and "likely to jeopardize"—as Steps 1, 2, and 3 in this report.

The committee does not take a position on any legal or regulatory policy issue, provide any legal or policy advice, or comment on the merit of any particular court ruling or other legal or policy decision. Furthermore, it recognizes that the agencies must make regulatory policy choices, and it has consciously avoided commenting on regulatory policy. In fact, the committee concludes that science and regulatory policy need to be kept separate to the extent possible and that there should be transparency where policy is involved. The present report evaluates the science of ecological risk assessment. Once an assessment is conducted, the involved agencies are responsible for making policy decisions pursuant to their legal mandates. The committee uses the generic term *decisionmaker* to indicate a person who will use the results of a risk assessment to inform a decision. The committee makes no statements on who such a person should be; that is a policy issue.

> **BOX 1-1** Statement of Task
>
> A committee of the National Research Council (NRC) will examine scientific and technical issues related to the methods and assumptions used by the U.S. Environmental Protection Agency (EPA), the U.S. Fish and Wildlife Service (FWS), and the National Oceanic and Atmospheric Administration (NOAA) to conduct scientific assessments of ecological risks from pesticides registered by EPA under the Federal Insecticide, Fungicide, and Rodenticide Act (FIFRA) to species listed under the Endangered Species Act (ESA). The range of scientific studies needed to make such assessments will be considered, including ecological, hydrological, toxicological, and exposure studies. The committee will develop conclusions reflecting the use of scientific principles and to facilitate a more holistic approach to assessing risks across the agencies, considering the intent of the ESA and of FIFRA. Policy issues related to decision making will not be addressed. Specific topics that the committee will consider to the extent practicable include the following:
>
> - **Best available scientific data and information**. The Services and EPA approach the identification of "best available scientific information" using a variety of differing protocols pertaining to the type and character of scientific information that may be appropriate for these evaluations. Some of these approaches pertain to the character of the information as consensus information, peer-reviewed information, regulatory studies supporting pesticide registrations, or other published and unpublished information. The NRC will evaluate those protocols with respect to validity, availability, consistency, clarity, and utility.
> - **Sublethal, indirect, and cumulative effects**. The ESA requires the consideration of direct, indirect, and cumulative effects on listed species and habitats in the consultation process. The Services and EPA have used differing approaches on how to characterize indirect, sub-lethal, and cumulative effects. The NRC will review the best available scientific methods for projecting these types of effects and consider options for the development of any additional methods that are likely to be helpful.
> - **Mixtures and inerts**. Assessing the effects of the use of chemical mixtures, either in formulated products or as used at the field level, remains a complex and difficult challenge, as is assessing the effects of mixtures of pesticides and other environmental contaminants. Projecting the effects of inert ingredients such as adjuvants, surfactants, and other pesticide product additives is also an area of continuing challenge. The NRC will consider the scientific information available to assess the potential effects of mixtures and inert ingredients.
> - **Models**. There is a range of approaches to the development and use of modeling to assist in analyzing the effects of actions such as using pesticides or alternatives to that use, and active issues remain about the use of unpublished models or the assumptions used in the choice of the available models for any particular analysis of effects. The NRC will assess the protocols governing the development of assumptions associated with model inputs and the use of sensitivity analyses to evaluate the impact of multiple assumptions on the interpretation of model results.

- **Interpretation of uncertainty.** There are a variety of methods for documenting and interpreting uncertainties and evaluating the extent to which uncertainties impact confidence in the scientific conclusions associated with a jeopardy decision. In particular, the NRC will consider the selection and use of uncertainty factors to account for lack of data on formulation toxicity, synergy, additivity, etc., and how the choice of those factors affects the estimates of uncertainty.
- **Geospatial information and datasets.** Location of the habitat is an important component of successfully protecting the impacted species. Much variability in datasets, geospatial layers, and scale contributes to uncertainty. The NRC will consider what constitutes authoritative geospatial information, including spatial and temporal scale that most appropriately delineates habitat of the species and the duration of potential effects.

In its deliberations, the NRC will focus on the scientific and technical methods and approaches the agencies use in determining risks to endangered and threatened species associated with the use of pesticides. The NRC will provide conclusions as appropriate about techniques the agencies might apply or use to improve those methods and approaches using scientific principles to support their decision-making.

As examples, the NRC will consider three recent consultations between NOAA and EPA on the effects of EPA's proposed FIFRA actions on Pacific salmonids as reference points for its work. The NRC will use the consultations as examples of the various agencies' scientific approaches and methods but will not evaluate the consultations themselves or the decisions resulting from them, and it will not limit its considerations strictly to aquatic species.

ORGANIZATION OF THE REPORT

The committee's report is organized into five chapters. Chapter 2 presents a common approach to the assessment process and discusses some overarching issues regarding uncertainty and best data available. Chapters 3 and 4 focus on exposure and effects analysis, respectively; each describes models and issues associated with uncertainty. Chapter 5 addresses the risk characterization process, which combines the results of the exposure and effects analyses. Excerpts of CFR Part 158 are provided in Appendix A, and Appendix B presents biographical information on the committee.

REFERENCES

Brennan, M.J., D.E. Roth, M.D. Feldman, and A.R. Greene. 2003. The Endangered Species Act: Thirty years of politics, money, and science. 387 square pegs and round holes: Application of the "best scientific data available" standard in the Endangered Species Act. Tulane Environ. Law J. 16(Sumer):387-444.

Doremus, H. 2004. The purposes, effects, and future of the Endangered Species Act's best available science mandate. Environ. Law 34:397-450.

2

A Common Approach and Other Overarching Issues

The committee was asked to comment specifically on scientific and technical approaches that might assist the US Environmental Protection Agency (EPA), the National Marine Fisheries Service (NMFS), and the Fish and Wildlife Service (FWS) in estimating risk to species listed under the Endangered Species Act (ESA) posed by pesticides (chemical stressors) under review by EPA for registration or reregistration as required by the Federal Insecticide, Fungicide, and Rodenticide Act (FIFRA). In this chapter, the committee discusses how the risk-assessment paradigm could serve as a common approach for EPA and the Services (NMFS and FWS) in examining the potential for listed species to be exposed to pesticides and the probability (that is, the risk) that such exposures would result in adverse effects. The risk-assessment paradigm was originally set forth in the report *Risk Assessment in the Federal Government: Managing the Process* (NRC 1983) and has been used and refined over the last few decades to evaluate both human health and environmental risks. Because this report is focused on risk to listed species in the environment posed by pesticide exposure, the committee focuses on ecological risk assessment (ERA) as described by such comprehensive references as Suter (2007). This chapter also addresses two general issues related to risk assessment: analysis of uncertainty and use of best data available.

A COMMON APPROACH

To comply with or administer the ESA during the pesticide registration process, EPA and the Services need to determine the probability of adverse effects on listed species or their habitats due to expected pesticide use that is consistent with label requirements. The committee understands that EPA and the Services are responding to different federal regulations and legal requirements and that the ESA places different responsibilities on the action agency (EPA) and the decision agency (NMFS or FWS). However, the committee has concluded that when the determination involves risk posed by chemical stressors, the agencies should use the same ERA paradigm to reach conclusions about adverse effects. Scientific obstacles to reaching agreement between EPA and the Services during consultation have emerged apparently because of the agencies' differences in implementation of the ERA process, including differences in underlying assumptions, technical approaches, data use, exposure models, and risk-calculation methods. Agreement has also been impeded because of a lack of communication and coordination throughout the process.

To understand and reconcile the differences between how EPA assesses risk to listed species from pesticide use and how the Services reach jeopardy decisions, it is important first to understand the consultation process under the ESA. The Services' *Endangered Species Consultation Handbook* (FWS/NMFS 1998) details the procedural and legal steps that they must follow when engaging in informal or formal consultations regarding listed species. As discussed in Chapter 1, the process involves three steps; the first two steps are to determine whether a proposed action needs formal consultation (Figure 2-1). In Step 1, the action agency (EPA) determines whether the action "may affect" a listed species. If the answer is yes (as it almost always is at the screening level for outdoor-use pesticides because "may affect" is interpreted broadly), EPA has two options: it can enter into formal consultation or proceed to Step 2—an optional step known as informal consultation—in which it must determine whether the action is "likely to adversely affect" a listed species. If the answer is no and NMFS or FWS concurs, the consultation process ends. However, if the answer is yes, Step 3 (formal consultation) is triggered. In formal consultation, NMFS or FWS must determine whether the action is "likely to jeopardize the continued existence of the species." A jeopardy decision must be informed by science, but the final regulatory determination of whether a risk is sufficient to constitute jeopardy is partly a policy decision. As the action agency, EPA is responsible for Step 1. It is also responsible, with concurrence from the Services, for Step 2; the Services are responsible for Step 3. In 2004, the Services promulgated a rule that would essentially authorize EPA to conduct Step 2 on its own without concurrence from the Services. The court found that this was a violation of the ESA, and it invalidated that portion of the Services' rule [*Washington Toxics Coalition v U.S. Fish & Wildlife Serv.*, 475 F.Supp.2d 1158 (W.D. Wash. 2006)]. In recent years, EPA seems to be bypassing Step 2 and initiating formal consultation whenever it finds that a pesticide "may affect" a listed species. Although this approach is permissible, it might be more efficient in many cases to conduct a Step 2 analysis before deciding to enter formal consultation. Presumably, Step 2 would filter out some actions, and fewer biological opinions would be needed. An agreed-on common approach to ERAs would give the Services more confidence in EPA's Step 2 analyses.

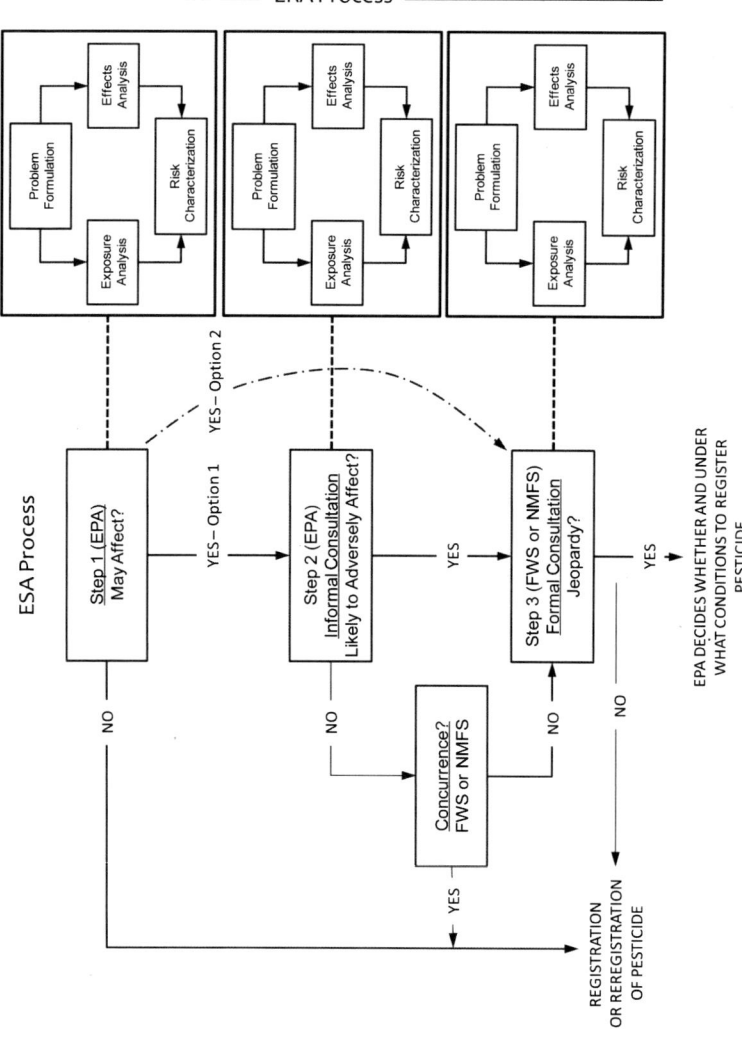

FIGURE 2-1 Relationship between the Endangered Species Act (ESA) Section 7 decision process and the ecological risk assessment (ERA) process for a chemical stressor. Each step answers the question that appears in the box.

As shown in Figure 2-1 and summarized in Table 2-1, the committee is suggesting that each step in an ESA consultation process for a chemical stressor be coordinated with an ERA process. Although the complexity of each ERA would depend on the step, each would involve the same four basic elements—problem formulation, exposure analysis, effects (or exposure-response) analysis, and risk characterization—that make up a risk assessment of a chemical stressor, such as a pesticide. The four basic elements and their relationships to one another trace their origin to the seminal *Risk Assessment in the Federal Government: Managing the Process* (NRC 1983; commonly referred to as the Red Book) and, more recently, to *Science and Decisions: Advancing Risk Assessment* (NRC 2009; commonly called the Silver Book). After 30 years of use and refinement, this risk-assessment paradigm has become scientifically credible, transparent, and consistent; can be reliably anticipated by all parties involved in decisions regarding pesticide use; and clearly articulates where scientific judgment is required and the bounds within which such judgment can be made. That process is used for human-health and ecological risk assessments and is used broadly throughout the federal government (for example, by the Food and Drug Administration). The committee notes that the Services' *Consultation Handbook* is silent regarding technical approaches to assessing risks to listed species posed by chemical stressors, such as pesticides. Consequently, the committee has concluded that the risk-assessment paradigm reflected in the ERA process is singularly appropriate for evaluating risks to ecological receptors, such as listed species, posed by chemical stressors, such as pesticides.

TABLE 2-1 Steps in the ESA Process as Related to Elements in the ERA Process for Pesticides[a]

Step in ESA Process [Responsible Agency]	Element of the ERA Process		
	Exposure Analysis (Chapter 3)	Effect (Exposure-Response) Analysis (Chapter 4)	Risk Characterization (Chapter 5)
1 [EPA] Determine whether use of a pesticide "may affect" any listed species	Distribution of listed species in space and time	Distribution of the pesticide in space and time if used as labeled (toxicity is assumed)	Possibility that species and pesticide distributions would overlap in space and time
2 [EPA] Determine whether use of a pesticide is "likely to adversely affect" any listed species	Modeled exposure concentrations	Exposure-response function for an individual receptor's survival and reproduction	Probability of adverse effects on survival and reproduction of individual receptors
3 [SERVICES] Determine whether use of a pesticide is likely to cause "jeopardy"	Modeled or measured exposure concentrations	Exposure-response functions for survival and reproduction rates	Probability of adverse effects on population viability over space and time

[a]See section "Coordination among Agencies" for a discussion of problem formulation, the first element of the ERA process.

Although the ERA process should always include the four elements, the content of each is expected to change as the question shifts from whether a pesticide "may affect" a listed species (Step 1) to whether it is "likely to affect" a listed species (Step 2) to whether the continued existence of the listed species is likely to be jeopardized (Step 3). Consistency of the basic ERA process throughout the three steps (if all are needed) is the first essential point. The second is that each ERA becomes more focused and specific to the chemicals and species of concern as it moves from Step 1 to Step 3. The third point is that the Services should build in Step 3 on what EPA did in Steps 1 and 2; they should not start over with a separate and different analysis in Step 3.

Thus, the committee envisions the following process. In Step 1, EPA would consider whether *any* listed species might be harmed by the pesticide simply by asking whether areas proposed for pesticide application and known (or suspected) species ranges or habitats coexist. Not all listed species exist everywhere, nor are all pesticides used everywhere, so that simple formulation of the problem would help to narrow the scope of later assessments. In Step 2, EPA would address the question of whether the use of a pesticide in the specific context of its proposed patterns of use is "likely to adversely affect" one or more listed species or their critical habitats. EPA would approach that question from a chemocentric viewpoint and estimate potential environmental concentrations and possible toxic effects. Essentially, EPA would evaluate whether the pesticide would be used in a manner that would result in environmental concentrations that have the potential to affect a listed species, other organisms in its habitat, or its critical habitat adversely. The assessment would be relatively generic (that is, not site-specific), and the effects analysis would focus on individuals of the listed species. If the predicted concentrations could adversely affect individuals in a population of a listed species, EPA would consult with the appropriate Service, which would then be responsible for a jeopardy determination. In Step 3, NMFS or FWS ideally would focus more specifically on potentially affected listed species in an ecological context and address the question of whether locally applicable predicted or measured exposures result in effects on the listed species or on other species in their habitats in a manner that would change the ability of a population to persist or to recover or that would change the time to extinction.

The possible differences in risk assessments between Steps 2 and 3 in the ESA process can be seen by considering that the imaginary pesticide X—designated PX for this discussion—will be applied to wheat in Illinois in summer. In this hypothetical example, EPA decides that because PX is used in a region where there are listed sturgeon species, some PX could get into streams and possibly affect the fish. So, the agency progresses to Step 2. Here, problem formulation is used to narrow the assessment's scope by asking two questions: Are there any organisms for which we know that the pesticide is nontoxic (for example, exposures at greater than 5,000 ppm cause no effect)? Are there any environmental media (water, soil, and air) in which the pesticide will not reside? Following problem formulation, EPA runs the standard farm-pond model to determine an initial estimate of the probable concentration of PX in the water, recognizing that the farm-pond model might not accurately represent conditions that apply to flowing streams and rivers where sturgeon actually live. That concentration is compared with the toxicity threshold that is based on full life-cycle tests in standard laboratory species, and EPA also considers that sturgeon are generally more sensitive to PX-like chemicals for the assessment end point than are standard test species. EPA concludes that pesticide concentrations in streams could exceed toxicity thresholds at the proposed application rates and notes further that PX-like chemicals can cause sublethal effects, including behavioral changes and darker color in adults. PX also kills aquatic invertebrates (the prey base of sturgeon) at concentrations lower than ones that affect sturgeon. On the basis of those results, EPA reaches a conclusion of "likely to adversely affect" and institutes formal consultation with the Services as required.

FWS builds on EPA's analysis in Step 3 and uses site-specific data on Illinois soils to calculate potential runoff to the slow-moving rivers and streams favored by sturgeon during summer. Because of the high clay content of the soils, PX binds to the root zone, and little is expected to move through soil into the streams. However, surface runoff—particularly during heavy rains, when a lot of soil is lost from fields—can result in water concentrations above effect concentrations. FWS reviews the information on behavioral effects and concludes that the studies are not reliable indicators of field effects. It also concludes that a darker color induced by PX exposure would increase the probability of survival of the fish because they would be more mud-colored and therefore better camouflaged. Because of concern about potential effects of PX on sturgeon in areas of the state that have a potential for substantial soil loss during summer rain events, FWS runs a spatially explicit population model to determine whether there could be a reduction in reproductive output that would affect the recovery of the population; it determines that changes in the growth rate of the population are unlikely. Furthermore, FWS concludes that the effects on aquatic invertebrates occur during times of the year when young sturgeon (the insectivorous life stage) are not present. Therefore, FWS reaches a conclusion of "no jeopardy."

In that hypothetical example, EPA and FWS use the same exposure models but different input parameters (generic farm-pond analyses vs site-specific soil runoff into shallow streams), assume different environmental transport pathways (surface runoff vs below ground), incorporate effects thresholds from the same studies, and review the same studies on sublethal effects. EPA uses reasonable worst-case assumptions of effects of PX on individual fish to reach a "likely to adversely affect" conclusion, whereas FWS uses site-specific data, incorporates spatial variability, and bases its decision on changes in population growth rates to reach a finding of "no jeopardy."

The committee concludes that using a common approach would eliminate many problems in assessing risks to listed species that are being encountered by EPA and the Services. As noted by Suter (2007, p. 37), the "advantages of using a single standard framework include familiarity and consistency, which reduce confusion and allow comparison and quality assurance of assessments." The ERA process that has evolved over the decades is best suited to evaluating the risk to listed species and their critical habitats posed by pesticides, and, as noted by Suter (2007, p. 37), the "EPA framework is a preferred default for ecological risk assessment in the United States." Although the committee does not expect the basic risk-assessment framework to change, it recognizes that risk-assessment approaches and methods for determining, for example, what is hazardous, how much is hazardous, what end points constitute an adverse effect, and when, where, and how much exposure is occurring will continue to evolve.

COORDINATION AMONG AGENCIES

A letter from the Services to EPA in 2004 (Williams and Hogarth 2004) detailed previous efforts to reconcile the differences between EPA's and the Services' approaches to pesticide evaluation. That letter was followed, in the same year, by an alternative consultation agreement between EPA and the Services. Although all six tasks assigned to the committee were discussed in that letter and the later agreement, the extent to which the agreement was implemented remains unclear. The committee emphasizes that given the changing scope of the ERA process from Steps 1-3, EPA and the Services need to coordinate to ensure that their own technical needs are met.

First, before a risk assessment is even initiated, the agencies need to connect the decision that must be made with the risk assessment that will inform it. That stage, often referred to as planning and scoping (EPA 1998, 2004), involves a team of decision-makers, stakeholders, and risk assessors who identify the problem to be assessed, develop a common understanding of why the risk assessment is being conducted, and establish the management goals of the assessment. Decision-makers can identify information that they need to make decisions, and risk assessors can ensure that the science meets the needs of decision-makers and stakeholders. Together, all stakeholders should be able to evaluate whether the assessment is likely to address the identified problems with the desired confidence (EPA 2004).

Second, problem formulation, conducted as part of the ERA process (see Figure 2-1), could provide an effective means for EPA and the Services to coordinate and reach agreement on many of the key technical issues involved in assessing risk posed to listed species by pesticides. Problem formulation frames the risk-management objectives sufficiently for the risk assessor to identify all potential inputs into the risk-assessment model. Guided by the needs of the decision-makers and using the best data available, the risk assessor develops a conceptual model of stressor sources, exposure pathways, and receptors; poses risk questions or hypotheses; and identifies the methods and analyses that will be used to address the questions and hypotheses. If problem formulation is successful, a comprehensive, scientifically credible conceptual model will be developed, there will be agreement on the risk-assessment approach, and the output of the assessment will have sufficient specificity for decision-making. The analysis phase of the risk assessment should not begin until the decision-makers are satisfied that the risk assessor understands the questions that need to be addressed and understands how much confidence in the final risk estimate is needed. Problem formulation is also an excellent time to discuss how the risk estimate will be communicated at the conclusion of the assessment.

The committee views coordination among EPA and the Services as a collegial exchange of technical and scientific information for the purpose of producing a more complete and representative assessment of risk, including the types and depths of analyses to be conducted at each step in the process. Such coordination would allow EPA's expertise in pesticides to be effectively combined with the Services' expertise in life histories of listed species and in abiotic and biotic stressors of the species. Coordination discussions would include many of the issues discussed by the committee in the present report, such as datasets to use to delineate species' habitats, the need for additional fate data, and new approaches for exposure and effects analysis. The agencies can use Steps 1-3 as a framework for such discussions but need not be constrained by them. It might be that technical working groups would form around various aspects of the assessment approach—such as fate and transport modeling, estimating species distributions and habitats, data-sharing, and uncertainty analysis—to discuss technical details and that others would discuss policy-based issues, such as which evolutionarily significant units to include in the analysis. The committee recommends that such collaboration meetings be formal, structured workshops that have stated goals and objectives, be led by professional facilitators, and have formal agendas agreed to by all parties. That approach would enhance productivity and allow expectations to be met. The periodicity of such discussions would necessarily be at the discretion of the agencies, but the committee recommends a frequency of at least once every 2 years to capture updates in risk-assessment and population-biology methods, newly listed species, new pesticide classes, and changing agricultural practices.

The committee concludes further that coordination during problem formulation regarding the ESA and ERA processes would be enhanced if a common outline, such as the one shown in Box 2-1, were adopted. The details of the outline would be adapted according to the step being conducted. However, the outline should incorporate specific elements of concern and interest to EPA and the Services. For example, examination of earlier EPA assessments has revealed a need for EPA to include and consider all available information about the life history of a listed species early in the process, ideally during planning and scoping (Item 1.1.4 in Box 2-1). Although assessment end points might ultimately involve only common surrogate or test species, the inclusion of natural life-history information on the listed species and critical habitat would at least enable a qualitative assessment of the similarities and differences between the listed species and the identified surrogates.

BOX 2-1 Generic Outline for Reporting Ecological Risk-Assessment Results for Listed Species or Their Critical Habitats

1. PROBLEM FORMULATION
   1.1. Background
      1.1.1. Defining the Regulatory Action
      1.1.2. Nature of the Pesticide
      1.1.3. Pesticide-Use Characterization
      1.1.4. Natural History of Listed Species
      1.1.5. Designated Critical Habitats
   1.2. Action Area (based on use and natural history)
   1.3. Assessment End Points
      1.3.1. Individuals
      1.3.2. Populations
      1.3.3. Critical Habitats
   1.4. Conceptual Model
      1.4.1. Risk Questions or Hypotheses
      1.4.2. Graphical Representation
   1.5. Analysis Plan
      1.5.1. Measures (exposure, effect, and characteristics)
      1.5.2. Approach to Risk Estimation

2. EXPOSURE ANALYSIS
   2.1. Label Application Rates and Intervals
   2.2. Habitats of Listed Species
   2.3. Exposure (Transport and Fate) Modeling
      2.3.1. Aquatic Organisms
      2.3.2. Terrestrial Organisms
   2.4. Exposure to Mixtures
   2.5. Monitoring Data
   2.6. Exposure Estimate (with uncertainty)

3. EFFECTS ANALYSIS
   3.1. Incident Database Review
   3.2. Individuals
      3.2.1. Direct Effects (acute, sublethal, and chronic)
      3.2.2. Indirect Effects
   3.3. Effects on Critical Habitats
   3.4. Mixture Effects
   3.5. Exposure-Response Estimate (with uncertainty)

4. RISK CHARACTERIZATION
   4.1. Risk Estimate
      4.1.1. Individuals
      4.1.2. Populations
      4.1.3. Critical Habitat
   4.2. Field and Laboratory Comparisons
   4.3. Risk Description (integration and synthesis)

UNCERTAINTY

The committee was asked to consider the interpretation of uncertainty and specifically the selection and use of uncertainty factors to account for lack of data. However, before the committee can answer the question about uncertainty factors, it must consider how uncertainty has been treated in past assessments. The committee addresses the question about uncertainty factors in Chapters 4 and 5.

In the context of this report, risk is defined as the probability of adverse effects on listed species or their critical habitats due to anticipated pesticide use that is consistent with label requirements. Ultimately, the adverse effect is jeopardy to the continued existence of a listed species defined in terms of demography, habitat, or other resources. The risk is estimated on the basis of predicted future pesticide exposure concentrations and the type and magnitude of effects (as determined by exposure-response functions) that the pesticide could have on the species. The risk estimate reflects uncertainty due to natural variability, lack of knowledge, and measurement and model errors in the host of underlying assumptions and variables used to predict exposure and effects. Natural variability or variation is true heterogeneity that might be better defined (but never eliminated) through increased sampling. Lack of knowledge (ignorance) is due to an absence of data or incomplete knowledge of important variables or their relationships; it can be reduced through additional data collection or further research. As indicated in Box 2-1, uncertainty will need to be characterized in the exposure estimation (Item 2.6) and the effect-response estimation (Item 3.5) analyses, then propagated, and finally integrated (Item 4.3) to provide the risk as a probability with an estimate of uncertainty.

The committee has concluded that achieving such integration will require that the ERA process in Steps 2 and 3 adopt a probabilistic approach that allows uncertainty in exposure and effect to be explicitly recognized and then combined to yield a risk as a probability with associated uncertainty (see Chapter 5). The present practice of relegating the consideration of uncertainty to a separate, often qualitative, narrative at the end of an assessment is of marginal value because doing so has little notable effect on risk estimation itself or on a decision-maker's ability to understand the confidence that should be placed in a risk estimate. Although the committee is aware of the administrative and other nonscientific hurdles that will need to be overcome to implement such an approach, it nonetheless has concluded that moving the uncertainty analysis from a narrative addendum to an integral part of the assessment is both possible and necessary to provide realistic, objective estimates of risk. Because a core dataset is required for all pesticide registration decisions, there should be sufficient information to conduct a quantitative assessment, which can include a quantification of the associated uncertainty.

The committee recognizes that the quantitative propagation of uncertainty through ecological risk assessments is not a new concept, particularly in the context of pesticide assessments. The topic was addressed by EPA's Scientific Advisory Panel for FIFRA in 1996 (Bailey et al. 1997) and was explicitly addressed in a workshop held in 2009 (Warren-Hicks and Hart 2010). EPA has since developed and begun to implement the Terrestrial Investigation Model (TIM; Odenkirchen 2003); TIM version 2.0 includes Monte Carlo simulations for calculating pesticide concentrations in a simulated farm pond and estimating activity patterns of potentially exposed wildlife. The committee recognizes that the use of frequentist statistics and Monte Carlo simulations, although widespread, is only one approach to quantifying and propagating uncertainty through an ERA. Bayesian approaches to environmental assessments, some of which also use Monte Carlo simulations, have become more widely understood and more feasible over the last few decades as computational power and capability have improved (Ellison 1996; McCarthy 2007; Link and Barker 2010). For example, Borsuk and Lee (2009) describe the application of Bayesian approaches to increase environmental realism in population modeling, and Reckhow (1999) applies similar approaches to water-quality predictions. Their applicability to analyses of data on chemicals and to other environmental risk assessments (Clark 2005), including those for endangered species, has been recognized in the federal government (FDA 2010; Conn and Silber 2013), although they have not yet been widely adopted for chemical risk assessment. Bayesian methods reliably estimate modeled variables, and Bayesian models can readily propagate uncertainties in data (such as measurement errors) and uncertainties in model structure (such as selection of covariates and relationships among them). The models can incorporate data from multiple sources, expert knowledge, and empirical evidence about relationships among variables and about the shape of the data distributions; however, these are not required to use or run the models. Bayesian approaches are most useful during Step 3 of ESA pesticide analyses when an in-depth analysis is needed, such as when alternative pesticide-use scenarios or proposed mitigation actions might have large spatial or economic consequences.

EPA has noted that "the explicit treatment of uncertainty during problem formulation is particularly important because it will have repercussions throughout the remainder of the assessment" (EPA 1998, p. 26). For ESA Section 7 consultations on pesticide risk to listed species, it is likely that the amount of data available for producing a risk estimate will vary by species and by chemical. The risk assessor will therefore need to ascertain during problem formulation how much confidence in the risk estimate the decision-maker requires to support a decision, given the decision context. Does the decision-maker need a risk estimate with low uncertainty or is, for example, ± 25% acceptable? Decisions regarding uncertainty need to be balanced with a discussion about availability of time and resources and need to consider the extent to which uncertainties are unavoidable given likely data gaps. A quantitative analysis of expected value of information could be conducted to answer the question of whether the reduction in uncertainty warrants obtaining more information (Yokota and Thompson 2004; Runge et al. 2011; Moore and Runge 2012). However, the committee recognizes that time limitations might preclude such an analysis (Yokota and Thompson 2004). The committee acknowledges the utility of a qualitative assessment and discussion between risk assessors and decision-makers at both Step 2 and Step 3 of the ESA risk-assessment process. A decision-maker will then be adequately informed about the estimated probability of an adverse effect and can make a decision about whether the proposed action is "likely to adversely affect" or can be "reasonably expected" to result in jeopardy. Decisions about the acceptable level of risk and how to manage the risk are policy decisions that are not part of the scientific analysis.

The committee recognizes that decision-makers will need to understand how to interpret and use the information on uncertainty in their decision-making. There is a great body of literature on risk management and decision-making under uncertainty that can help to guide and guard against misuses of uncertainty in decision-making (see, for example, Cropper et al. 1992, Morgan and Henrion 1992, EPA 2010, and IOM 2013).

BEST DATA AVAILABLE

As discussed in Chapter 1, the Services have a mandate to use the "best scientific and commercial data available" in their assessments. There is little guidance on what constitutes "best data available," and the agencies do not appear to have formal protocols that define "best data available." However, the following sections describe the agencies' approaches to data collection and evaluation, and the committee provides some guidance on important data characteristics.

Scope of Data Collection and Selection

EPA (1998) indicated that a search for all available data is conducted at the start of each risk assessment and iteratively throughout the assessment to support and guide each step of the process. EPA's primary repository for peer-reviewed toxicity studies that are publicly available is ECOTOXicology (ECOTOX; EPA 2012a). The Services and EPA agreed to use ECOTOX as the common source for data on ecotoxic effects of pesticides (EPA 2011).

Data used by EPA in pesticide risk assessments are typically derived from detailed reports of standardized studies required for pesticide registration under FIFRA; studies in peer-reviewed journals or other publications, such as reference books; and government reports and surveys. Repository databases are used if they meet data-quality standards. The Services also include anecdotal or oral information and other unpublished materials from such sources as state natural-resources agencies and natural-heritage programs, tribal governments, other federal agencies, consulting firms, contractors, and persons associated with professional organizations and institutions of higher education (59 Fed. Reg. 34275 [1994]). Accordingly, the scope of data collection by the Services appears broader, although some of the information collected can be brief and be insufficient for independent evaluation.

Evaluation of Data Relevance and Quality

Information on pesticides and the ecology of listed species that is used in risk assessments should be both relevant and of high quality. Relevance refers to information that is consistent with its intended use. Accordingly, the information should be from studies of the species and chemicals being assessed, or there should be a strong theoretical basis for extrapolation to the species and chemicals being assessed. The information should be spatially applicable to the locations being considered and be sufficiently recent to be pertinent. For example, information on the environmental transport and fate of the specific pesticide active ingredient under review or of the class to which the pesticide belongs would be relevant to the assessment. Similarly, information on the ecology of the listed species is highly relevant and useful particularly if it has been obtained recently from the area of pesticide use. Conversely, a study that used a population census of the listed species conducted 20 years ago would not be relevant. Information that is not relevant clearly should not be used to assess risk, and the question of relevance is the first question that needs to be addressed in considering whether information should be used for a risk assessment.

The quality of the relevant information should be reviewed before it is used in a risk assessment. A critical question to answer is whether the data conform to best scientific practice. Best practice includes providing sufficient information that characterizes the data (such as who collected them, when and where they were collected, what variables were measured, and how and in what units measurements were taken), clear methods that would allow a third party to replicate the data-collection process or the analyses conducted with the data, and estimates of data accuracy or uncertainty. If sufficient information is not available, data quality is unknown, and the data should be given less prominence in the risk assessment. Accordingly, data of lower quality should not be used to nullify data of higher quality. Ideally, data are objective and unbiased, although failure to meet those requirements might not be a cause for rejection if biases are sufficiently described and clearly identified in the assessment.

EPA has a formal set of data relevance and quality criteria that are applied in selecting information for use in regulatory assessment. The EPA Science Policy Council published a set of five assessment factors for evaluating scientific and technical information on the basis of EPA practices, input from the public, and results from a workshop hosted by the National Academy of Sciences (EPA 2003). The assessment factors are intended to improve data generation, use, and dissemination in EPA and by the data-generating public. The assessment factors are *applicability and utility* (relevance of the information to its intended use and applicability to the current scenarios of concern), *soundness* (scientific validity of experimental study, survey, modeling, and data collection and adequate support for conclusions), *clarity and completeness* (documentation that includes underlying assumptions, study protocol and design, data accessibility, and data analysis), *uncertainty and variability* (quantitative and qualitative characterization, effect on conclusions, and the identification of parameter values that, if changed, would substantially affect the outcome of the model), and *evaluation and review* (independent verification, validation, and peer review and consistency with results of similar studies). The EPA Office of Pesticide Programs has additional guidelines for acceptance of scientific literature, as described in the documentation supporting the ECOTOX database (EPA 2012b).

FWS and NMFS do not have agency-specific guidelines on data relevance and quality. However, all federal agencies are expected to comply with the Office of Management and Budget (OMB) guidelines on objectivity, utility, and integrity of disseminated information. OMB (67 Fed. Reg. 8452 [2002]) describes those attributes as follows:

> "Objectivity" focuses on the extent to which information is presented in an accurate, clear, complete and unbiased manner; and, as a matter of substance, the extent to which the information is accurate, reliable and unbiased. "Utility" refers to the usefulness of the information to the intended users. "Integrity" refers to security, such as the protection of information from unauthorized access or revision, to ensure the information is not compromised through corruption or falsification.

The Services and EPA (EPA 2002; FWS 2007) have separately published information quality guidelines (IQGs) that follow closely the government-wide OMB guidelines. Similar basic principles for achieving a scientifically credible assessment are prescribed in the IQGs from the agencies; the agencies are committed to ensuring the quality of evaluations and the transparency of information from external sources used in their disseminated assessments and actions (EPA 2003; NMFS 2005). They also recognize that a high level of transparency and scrutiny is needed for influential information that is expected to have a substantial effect on policies and decisions (EPA 2002; NMFS 2004; FWS 2007).

In the biological opinions provided, the committee was able to discern at least one approach that the Services use to evaluate relevance and quality of data. In the ESA consultation for assessing the effects of 12 organophosphates on salmonids (NMFS 2010), NMFS described and used a qualitative set of evaluation criteria. Three criteria were used to judge the relevance of the publicly available toxicity data: whether the studies were conducted on salmonids, whether they measured end points of concern, and whether they evaluated effects of exposure to the specific chemicals or structurally related chemicals. The more criteria were met, the more relevant the studies were deemed. A fourth criterion was related to data quality and had to do with whether relevant studies had substantial flaws in experimental design.

Important Data Characteristics

Data relevance and data quality clearly are primary factors in determining whether data constitute "best available data." Several data characteristics noted in the committee's charge and described below are related to relevance and quality and can help to determine whether data are useful for assessing the risk to listed species posed by pesticides.

Validity. Data that are used in risk assessment should be accompanied by sufficient information for repeatability, independent scientific review, and additional data analysis when needed (NRC 1995). For example, an additional analysis, such as a dose-response analysis, might be necessary to ensure accurate interpretation of the data. Data sources that lack sufficient details for an adequate scientific evaluation—such as poster presentations, abstracts, anecdotal or personal communications, and data files that contain no information on fundamental data attributes—might provide background knowledge or support an overall weight-of-evidence evaluation but should not be the sole basis for drawing assessment conclusions. Thus, although secondary information can be useful for identifying an original report, it should not be used directly in risk assessment. The original study is necessary for an independent review of accuracy, quality, and relevance. An example from the draft biological opinion on the effects of 2,4-dichlorophenoxyacetic acid (2,4-D) on salmonids illustrates the problems with using secondary sources. That biological opinion cited Brock et al. (2000), which attributed a value for an aquatic-community effect to a report by Boyle (1980), but the effect cited was not found in the primary source.

Availability. Many data used in pesticide risk assessment are taken from unpublished studies that are conducted to support pesticide registrations. Those studies are conducted according to well-defined protocols and prescribed good laboratory practices. The detailed reporting allows EPA scientists to evaluate study quality independently and to conduct data analysis beyond what is possible with studies published in the open literature.[1] EPA's evaluation is documented in a data evaluation record (DER), which contains information on study methods, results, and discussions. Additional data analysis or modeling is also documented. Recent DERs, in contrast with older ones, can serve as stand-alone reports based on full study reports submitted for pesticide registration. Public availability of DERs is important because the submitted studies are typically protected confidential business information (CBI) and not publicly available or readily accessible. However, other government agencies, such as NMFS and FWS, can review CBI once necessary information controls are in place and therefore provide data-quality assurance for EPA's reported information on in-

---

[1] As noted in Chapter 1, the Services do not have the authority under the ESA to require the generation of data but instead must rely on the best data that are available. Furthermore, the ESA makes it clear that the Services are not to delay action because of a lack of data.

dustry studies. In addition, EPA has increased public access to DERs in recent years by making more information available to the general public. The committee encourages EPA to continue to share the studies with the Services, to provide sufficient details in DERs to ensure a reasonable understanding of the studies, and to make DERs readily available to the public.

Consistency. Data consistency is an important consideration in drawing scientific inferences. Apparently conflicting results from different studies should be examined with care. Different results from studies that use different species, life stages, exposure regimens, observation methods, experimental conditions, or statistics do not necessarily constitute conflicting evidence, and all might be useful in drawing conclusions. However, statistical outliers should be given particular scrutiny to verify the quality of an underlying study, particularly if they differ from all other data by orders of magnitude.

Clarity. The strengths and weaknesses of data and the reason that they were or were not used in a risk assessment should be clearly documented. Expert opinion or judgment is also used in risk assessment and is valuable especially when uncertainty is high because of data gaps. However, it is important that the assumptions or judgments be clearly described. As stated in NRC (1995), a clear presentation of expert knowledge should include the line of reasoning used and should separate facts from speculation. Similarly, adequate rationale should be given throughout the assessment for the assumptions that are made in the absence of data.

Utility. Utility clearly is related to relevance. One specific issue that has arisen regarding utility concerns the usefulness of foreign-language articles. Studies might be excluded by EPA because of a language barrier and lack of funding to obtain an English translation. For example, foreign-language reports, especially ones that are not readily available in the open literature, might be included in ECOTOX but not used in a risk assessment. If foreign-language reports are used in a risk assessment, translated versions will be needed so that the data in the reports can be subjected to the same data quality and relevance evaluation as data from studies published in English.

Peer Review. Regardless of the data criteria, it is not unusual for well-qualified risk assessors to disagree on the quality of data or on their relevance to a specific assessment. Because OMB attaches stricter requirements to discretional peer review of highly influential scientific assessments (Bolten 2004), the committee emphasizes the value of external peer review to enhance the quality, transparency, and credibility of a risk assessment.

CONCLUSIONS AND RECOMMENDATIONS

A Common Approach and Coordination among the Agencies

- Lack of a common approach has created scientific obstacles to reaching agreement between EPA and the Services during consultation.

- The risk-assessment paradigm, as reflected in the ERA process, is a scientifically credible basis of a single, unified approach for evaluating risks to listed species posed by pesticide exposure under FIFRA and the ESA.
- The committee's recommendation is that the ERA process include the same four elements (problem formulation, exposure analysis, effects analysis, and risk characterization) at each step but that the content of each changes as the question shifts from whether the pesticide "may affect" a listed species (Step 1) to whether it is "likely to adversely affect" a listed species (Step 2) to whether the continued existence of the listed species is jeopardized (Step 3).
- The ERA process would be enhanced if it were accompanied by use of a common outline that incorporates specific elements of concern to EPA and the Services.
- Given the changing scope of the ERA process from Step 1 to Step 3, EPA and the Services should coordinate to ensure that their own technical needs are met.
- Problem formulation, conducted as part of the ERA process, could be an effective way for the agencies to coordinate and reach agreement on many of the key technical issues involved in assessing risks posed by pesticide exposure.

Uncertainty

- Risk assessments and jeopardy decisions require recognizing and analyzing uncertainty and quantitatively propagating it through any assessment so that it is clearly reflected in the eventual risk estimate.
- The agencies should adopt a probabilistic approach that allows uncertainty in exposure and effect to be explicitly recognized and then combined in forming a risk estimate.
- Although administrative and other nonscientific hurdles will need to be overcome to implement such an approach, changing uncertainty analysis from a narrative addendum to an integral part of the assessment is possible and necessary to provide realistic, objective estimates of risk.
- Decisions about acceptable levels of risk and how to manage risk are policy decisions that are not part of the scientific analysis.

Best Data Available

- The agencies do not appear to have formal protocols for defining "best data available" and appear to approach data collection and selection from different perspectives.
- To ensure that the best data available are captured, a broad data search is needed at the beginning of the process. Dates of searches and search strategies should be clearly documented to ensure transparency of the process. If a repository database is searched, its contents and scope should be described, including

criteria for data inclusion and exclusion, periodicity of updates, and quality control for data entry.

- Given that stakeholders are aware of and can provide valuable and relevant data, the committee encourages provision for their involvement at the early stage and throughout the ERA process. Stakeholder data are expected to meet the same data relevance and quality standards as all other data.
- To ensure that the best data available are used, information should first be screened for relevance and then subjected to quality review.
- The agencies should, at a minimum, subject all information to a review based on OMB criteria of "objectivity, utility and integrity." Information sources that fail any of the criteria can be used at the discretion of the risk assessor, provided that their limitations are clearly described.
- Comparisons of all information sources with the relevance and quality attributes should be documented in the risk assessment and described in the overall characterization of uncertainties.

REFERENCES

Aguilera, P.A., A. Fernadez, R. Fernandez, R. Rumi, and A. Salmeron. 2011. Bayesian networks in environmental modelling. Environ. Modell. Softw. 26(12):1376-1388.

Bailey, T., E. Fite, P. Mastradone, B. Montague, R. Parker, I. Sunzenauer, and J. Wolf. 1997. EFED's Summary of the SAP Review of Ecological Risk Assessment Methodologies. April 14, 1997 [online]. Available: http://www.epa.gov/oppefed1/ecorisk/ecofram/efedsum.htm [accessed Feb. 22, 2013].

Bolten, J.B. 2004. Issues of OMB's "Final Information Quality Bulletin for Peer Review". Memorandum for Heads of Departments and Agencies, from Joshua B. Bolten, Director, Office of Management and Budget, Washington, DC. M-05-03, December 16, 2004 [online]. Available: http://www.whitehouse.gov/sites/default/files/omb/memoranda/fy2005/m05-03.pdf [accessed Nov. 9, 2012].

Borsuk, M.E., and D.C. Lee. 2009. Stochastic population dynamic models as probability networks. Pp. 199-219 in Handbook of Ecological Modeling and Informatics, S.E. Jorgensen, T.S. Chon, and F. Recknagel, eds. Billerica, MA: WIT Press.

Boyle, T.P. 1980. Effects of the aquatic herbicide 2, 4-D DMA on the ecology of experimental ponds. Environ. Pollut. A Ecol. Biol. 21(1):35-49.

Brock, T.C.M., J. Lahr, and P.J. Van den Brink. 2000. Ecological Risk of Pesticides in Freshwater Ecosystems, Part 1. Herbicides. Alterra-Rapport 088. Alterra, Green World Research, Wageningen, The Netherlands [online]. Available: http://edepot.wur.nl/18131 [accessed Nov. 5, 2012].

Charest, S. 2002. Bayesian approaches to the precautionary principal. Duke Env. L. Policy F. 12(2):265-291.

Clark, J.S. 2005. Why environmental scientists are becoming Bayesians. Ecol. Lett. 8:2-14.

Conn, P.B., and G.K. Silber. 2013. Vessel speed restrictions reduce risk of collision-related mortality for North Atlantic right whales. Ecosphere 4(4):Art. 43.

Cropper, M., W.N. Evans, S.J. Berardi, M.M. Ducla-Soares, and P.R. Portney. 1992. The Determinants of pesticide regulation: A statistical analysis of EPA decision-making. J. Polit. Econ. 100(1):175-197.

Ellison, A.M. 1996. An introduction to Bayesian inference for ecological research and environmental decision-making. Ecol. Appl. 6(4):1036-1046.

EPA (U.S. Environmental Protection Agency). 1998. Guidelines for Ecological Risk Assessment. EPA/630/R-95/002F. Risk Assessment Forum, U.S. Environmental Protection Agency, Washington, DC. April 1998 [online]. Available: http://www.epa.gov/raf/publications/pdfs/ECOTXTBX.PDF [accessed Nov. 5, 2012].

EPA (U.S. Environmental Protection Agency). 2002. Guidelines for Ensuring and Maximizing the Quality, Objectivity, Utility, and Integrity of Information Disseminated by the Environmental Protection Agency. EPA/260R-02-008. Office of Environmental Information, U.S. Environmental Protection Agency, Washington, DC. October 2002 [online]. Available: http://www.epa.gov/quality/informationguidelines/documents/EPA_InfoQualityGuidelines.pdf [accessed Nov. 5, 2012].

EPA (U.S. Environmental Protection Agency). 2003. A Summary of General Assessment Factors for Evaluating the Quality of Scientific and Technical Information. EPA 100/B-03/001. Science Policy Council, U.S. Environmental Protection Agency, Washington DC. June 2003 [online]. Available: http://www.epa.gov/stpc/pdfs/assess2.pdf [accessed Nov. 5, 2012].

EPA (U.S. Environmental Protection Agency). 2004. Overview of the Ecological Risk Assessment Process in the Office of Pesticide Programs, U.S. Environmental Protection Agency: Endangered and Threatened Species Effects Determinations. Office of Prevention, Pesticides and Toxic Substances, Office of Pesticide Programs, U.S. Environmental Protection Agency, Washington, DC. January 23, 2004 [online]. Available: http://www.epa.gov/espp/consultation/ecorisk-overview.pdf [accessed Nov. 5, 2012].

EPA (U.S. Environmental Protection Agency). 2010. Integrating Ecological Assessment and Decision-Making at EPA: A Path Forward. EPA/100/R-10/004. Risk Assessment Forum, U.S. Environmental Protection Agency, Washington, DC. December 2010 [online]. Available: http://www.epa.gov/raf/publications/pdfs/integrating-ecolog-assess-decision-making.pdf [accessed Mar.21, 2013].

EPA (U.S. Environmental Protection Agency). 2011. Procedures for Screening, Reviewing, and Using Published Open Literature Toxicity Data in Ecological Risk Assessments. Office of Pesticide Programs, U.S. Environmental Protection Agency. May 9, 2011 [online]. Available: http://www.epa.gov/pesticides/science/efed/policy_guidance/team_authors/endangered_species_reregistration_workgroup/esa_evaluation_open_literature.htm#guidance [accessed Nov. 5, 2012].

EPA (U.S. Environmental Protection Agency). 2012a. ECOTOX Database Release 4.0 [online]. Available: http://cfpub.epa.gov/ecotox/ [accessed Nov. 12, 2012].

EPA (U.S. Environmental Protection Agency). 2012b. ECOTOX Limitations. ECOTOX Database Release 4.0 [online]. Available: http://cfpub.epa.gov/ecotox/help.cfm?help_id=limitations [accessed Nov. 12, 2012].

FDA (Food and Drug Administration). 2010. Guidance for the Use of Bayesian Statistics in Medical Device Clinical Trials. Guidance for Industry and FDA Stuff. Food and Drug Administration, Center for Devices and Radiological Health, Division of Biostatistics, Washington DC. February 5, 2010 [online]. Available: http://www.fda.gov/downloads/MedicalDevices/DeviceRegulationandGuidance/GuidanceDocuments/ucm071121.pdf [accessed Feb.33, 2013].

FWS (U.S. Fish and Wildlife Service). 2007. Information Quality Guidelines. U.S. Fish and Wildlife Service [online]. Available: http://www.fws.gov/informationquality/topics/IQAguidelines-final82307.pdf [accessed Nov. 5, 2012].

FWS/NMFS (U.S. Fish and Wildlife Service and National Marine Fisheries Service). 1998. Endangered Species Consultation Handbook: Procedures for Conducting Consultation and Conference Activities Under Section 7 of the Endangered Species Act. U.S. Fish and Wildlife Service and National Marine Fisheries Service, Washington, DC [online]. Available: http://sero.nmfs.noaa.gov/pr/esa/pdf/Sec%207%20Handbook.pdf [accessed Nov. 5, 2012].

IOM (Institute of Medicine). 2013. Environmental Decisions in the Face of Uncertainty. Washington, DC: National Academies Press.

Link, W.A., and R.J. Barker. 2010. Bayesian Inference: With Ecological Applications. New York: Academic Press.

McCarthy, M.A. 2007. Bayesian Methods for Ecology. New York: Cambridge University Press.

Moore, J.L., and M.C. Runge. 2012. Combining structured decision making and value-of-information analyses to identify robust management strategies. Conserv. Biol. 26(5):810-820.

Morgan, M.G., and M. Henrion. 1992. Uncertainty: A Guide to Dealing with Uncertainty in Quantitative Risk and Policy Analysis. Cambridge: Cambridge University Press.

NMFS (National Marine Fisheries Service). 2004. Section 515 Pre-dissemination Review and Documentation Guidelines. Data Quality Act. National Marine Fisheries Service Policy Directive PD 04-108-03-2004. Renewed July 27, 2012 [online]. Available: http://www.nmfs.noaa.gov/op/pds/documents/04/108/04-108-03.pdf [Nov. 12, 2012].

NMFS (National Marine Fisheries Service). 2005. Policy on the Data Quality Act. Policy Directive PD 04-108, December 30, 2005. Renewed July 27, 2012 [online]. Available: http://www.nmfs.noaa.gov/op/pds/documents/04/04-108.pdf [accessed Nov. 12, 2012].

NMFS (National Marine Fisheries Service). 2010. Endangered Species Act Section 7 Consultation, Biological Opinion - Environmental Protection Agency Registration of Pesticides Containing Azinphos methyl, Bensulide, Dimethoate, Disulfoton, Ethoprop, Fenamiphos, Naled, Methamidophos, Methidathion, Methyl parathion, Phorate and Phosmet. August 31, 2010 [online]. Available: http://www.nmfs.noaa.gov/pr/pdfs/final_batch_3_opinion.pdf [accessed Nov. 5, 2012].

NRC (National Research Council). 1983. Risk Assessment in the Federal Government: Managing the Process. Washington, DC: National Academy Press.

NRC (National Research Council). 1995. Science and the Endangered Species Act. Washington, DC: National Academy Press.

NRC (National Research Council). 2009. Science and Decisions: Advancing Risk Assessment. Washington, DC: National Academies Press.

Odenkirchen, E. 2003. Evaluation of OPP's Terrestrial Investigation Model Software and Programming to Meet Technical/Regulatory Challenges. Presentation at EUropean FRamework for Probabilistic Risk Assessment of the Environmental Impacts of Pesticides Workshop, June 5-8, 2003, Bilthoven, the Netherlands [online]. Available: http://www.epa.gov/oppefed1/ecorisk/presentations/eufram_overview.htm [accessed Feb. 22, 2013].

Reckhow, K.H. 1999. Water quality prediction and probability network models. Can. J. Fish. Aquat. Sci. 56:1150-1158.

Runge, M.C., S.J. Converse, and J.E. Lyons. 2011. Which uncertainty? Using expert elicitation and expected value of information to design an adaptive program. Biol. Conserv. 144(4):1214-1223.Warren-Hicks, B., and A. Hart. 2010. Application of Uncertainty Analysis to Ecological Risks of Pesticides. Boca Raton: CRC Press.

Suter II, G.W. 2007. Ecological Risk Assessment, 2nd Ed. Boca Raton, FL: CRC Press. 643 pp.

Warren-Hicks, W.J., and A Hart, eds. 2010. Application of Uncertainty Analysis to Ecological Risks of Pesticides. Pensacola, FL: SETAC Press.

Williams, S., and W. Hogarth. 2004. Services Evaluation U.S. EPA's Risk Assessment Process. Letter to Susan B. Hazen, Principal Deputy Assistant Administrator, Office of Prevention, Pesticides and Toxic Substances, U.S. Environmental Protection Agency, from Steve Williams, Director, U.S. Fish and Wildlife Service, and William Hogarth, Assistant Administrator, National Marine Fisheries Service. January 24, 2004 [online]. Available: http://www.epa.gov/oppfead1/endanger/consultation/evaluation.pdf [accessed Nov. 12, 2012].

Yokota, F., and K.M. Thompson. 2004. Value of information analysis in environmental health risk management decisions: Past, present, and future. Risk Anal. 24 (3):635-650.

3

Exposure

The committee was asked to consider various issues associated with models, geospatial data, mixtures, and uncertainty. Although the language of the task statement was focused on effects analysis, determining which effects might be relevant requires estimating exposure. In this chapter, the committee first discusses fate and transport models used in exposure analyses by the agencies and then provides suggestions for a stepwise approach to estimating environmental concentrations of pesticides in the context of complying with the Endangered Species Act (ESA). Next, the committee addresses what constitutes authoritative geospatial data—critical information used to conduct exposure modeling and define species' habitats—and provides some examples. Finally, the committee discusses some important uncertainties associated with exposure analysis and the need to propagate uncertainty through the analysis.

EXPOSURE-MODELING PRACTICES

If pesticides are to be used without jeopardizing the survival of listed species and their habitats, the estimated environmental concentrations (EECs) to which the organisms and their habitats will be exposed need to be determined. Chemical fate and transport models are the chief tools used to accomplish that task. Broadly, such a model requires a user to choose a series of environmental control volumes—that is, environmental compartments containing multiple media, such as air, water, and soil—that are assumed to have a single, homogeneous pesticide concentration at each time step of the model. The transport and transformation processes that might affect a pesticide's presence in each control volume are combined and assembled into a mass-balance model that allows estimation of the EECs. Typically, the fate processes, such as sorption and biodegradation, are mathematically expressed in such a way that they can be adjusted by using chemical-specific and environment-specific information. However, knowledge or information can be insufficient, so the model parameter values for some chemical or physical processes are often oversimplified. For example, the distribution of a pesticide between the solids and water in a single compartment might be quantified by using a linear adsorption isotherm, although the data might suggest that the pesticide sorption mechanism exhibits nonlinear behavior.

Because the pathways by which pesticides move from their points of application to habitats of listed species might involve a complex sequence of transfers and diverse degradation processes, it is common to use a linked series of models to estimate exposure. Fate and transport modeling practices used by the US Environmental Protection Agency (EPA), Fish and Wildlife Service (FWS), and National Marine Fisheries Service (NMFS) are discussed below. The committee also elaborates on its suggestions for analyses that comply with Steps 1-3 in the ESA process when estimating exposure (see Table 2-1).

Approaches and Models Used by the Agencies

In Step 1 of the ESA process, EPA uses a program called DANGER to determine which listed species or their habitats coincide geographically and temporally with areas of pesticide use (EPA 2012a).[1] DANGER is an electronic database of county-level information on occurrence of listed species and acreage of agricultural crops. If there is geographic and temporal overlap, EPA assumes a "may affect" for pesticide use and addresses the listed species during its pesticide risk assessment (Step 2), in which pesticide concentrations are estimated in the environmental media to which the species might be exposed, as discussed below.

In Step 2 of the ESA process, EPA first uses a generic screening model to determine whether the pesticide is likely to move off the crop and into a body of water in concentrations high enough to trigger a concern for any aquatic species. For that initial screen, EPA uses GENEEC2 (Generic Estimated Environmental Concentration) (EPA 2001), a model that estimates pesticide concentrations in a standard small farm pond (a 2-m deep pond that has a surface area of 1 hectare in a watershed area of 10 hectares), uses generic inputs, and simulates a single event. Few fate processes are considered in the model. EPA typically assumes the maximum pesticide application rate as allowed by the label, and the model estimates pesticide concentration in the pond on the basis of spray drift and runoff from a 6-in. rain event that lasts 24 h.

As a screening model, GENEEC is sometimes characterized as providing worst-case estimates of exposure. The term *worst-case*, however, is misleading and should be avoided. The documentation for the model does not use the term *worst-case* but states that GENEEC "may provide a good predictor of upper level pesticide concentrations in small but ecologically important upland streams" (EPA 2001). That conclusion is attributed to Effland et al. (1999), but they discuss general monitoring data in streams rather than specific field studies that might be used to evaluate the accuracy of GENEEC with respect to specified applications.

---

[1]The committee understands that EPA now commonly refers to the DANGER database as LOCATES (A. Pease, EPA, personal commun., May 13, 2013).

Exposure

If the initial screening assessment triggers a concern for any aquatic species, EPA uses more sophisticated models, such as the Plant Root Zone Model (PRZM3; Suarez 2005) and the Exposure Analysis Modeling System (EXAMS; Burns 2004), to estimate pesticide concentrations in surface waters (EPA 2012b,c). Again, the standard farm field (10 hectares) and pond (1 hectare) scenario is typically modeled, but the models incorporate more fate processes and simulate effects of daily weather variability over multiple years. For example, the initial spatial fallout of a pesticide sprayed via aircraft into air over a field is estimated with a model, such as AgDRIFT® (Teske et al. 2002; SDTF 2010). The AgDRIFT-derived estimates then serve as inputs into PRZM3, which assesses pesticide fate in the soil environment, including evaporation to the atmosphere, infiltration into the subsurface, and off-site transport via overland runoff. Finally, to the extent that the combination of AgDRIFT and PRZM3 (which includes the Vadose Zone Flow and Transport model subroutine) yields estimates of pesticide delivery to nearby surface waters, EXAMS is used to estimate the temporally changing chemical concentrations in those waters and their underlying sediments. The resulting estimated concentrations in soil, water, and sediment yield estimates of the pesticide exposure of receptors of interest, including listed species.

For terrestrial species, EPA models pesticide exposure with the Terrestrial Residue Exposure (T-REX) model, the TerrPLant model, the Screening Imbibition Program (SIP) model, and the Screening Tool for Inhalation Risk (STIR) model (EPA 2012d). Exposure of terrestrial species is assumed to be through the diet, which is simulated by the exposure routine in T-REX. The model calculates pesticide residue concentrations on various food items (for example, short grass and broad-leafed plants) on the basis of work by Hoerger and Kenaga (1972) as modified by Fletcher et al. (1994) at a daily interval for 1 year. Other parts of the T-REX model translate exposure concentrations into daily doses for hypothetical small, medium, and large birds and mammals on the basis of food intake-rate equations from EPA's *Wildlife Exposure Factors Handbook* (EPA 1993). More recently, EPA has begun to estimate wildlife exposure through drinking water with the SIP model and inhalation with the STIR model. Those models are intended for use during problem formulation to determine whether the alternative exposure routes should be considered in the aggregate with food ingestion. SIP assumes that water concentrations are at the limit of solubility, and drinking-water ingestion rates are from Nagy and Peterson (1988). STIR calculates vapor-phase exposure from chemical-specific properties, such as molecular weight and vapor pressure, and includes estimates of spray-droplet exposure. Maximum inhalation rates are from EPA (1993), and the model assumes that a small-bodied bird or mammal is exposed to saturated air. For terrestrial plants, exposure for screening-level assessments of single pesticide applications is estimated by TerrPLant by assuming runoff delivery from a treated dry acre of land to a neighboring untreated acre, runoff from 10 treated acres to a 1-acre neighboring wetland, or specified percentages of spray drift after ground and aerial applications.

In Step 3, the Services also calculate environmental exposures with the same models that EPA uses in Step 2. For example, GENEEC2 was used in some of the biological opinions (BiOps) reviewed by the committee (NMFS 2008, pp. 235ff; 2009, pp. 284ff; 2010, pp. 294ff) as was AgDRIFT (NMFS 2008, p. 228). The committee did not examine any BiOps on terrestrial organisms, so it cannot comment on the terrestrial-exposure models used by the Services. However, the model input parameters used by NMFS to estimate aquatic exposure concentrations differ from those used by EPA, and the model is modified to estimate input into waters other than the standard farm pond. Those differences account for regional and habitat differences that are specific to the listed species and are discussed further in the next section.

A Stepwise Approach to Fate and Transport Modeling

Mass-balance models for chemical exposure analyses have several strengths. First, principles of mass-balance modeling and computer-simulation programs are well established. Second, many exposure models—such as AgDRIFT, PRZM, and EXAMS—are well documented. Third, the models can be made case-specific by time-varying data, such as meteorological conditions. Fourth, the output of one model can be used as input into the next one; for example, EXPRESS is a linked EXAMS-PRZM Exposure Simulation Shell (Burns 2006).

However, the model limitations need to be recognized, and models need to be used in the appropriate contexts. For example, GENEEC2 was developed by EPA simply as an easy-to-use screening tool to provide a consistent approach in the conduct of screening-level assessments, such as in Step 1 (or early in Step 2) of the ESA process (see Table 2-1). Although the Services have used GENEEC2 in BiOps, the committee concludes that a screening-level model has no place in Step 3 of the ESA process, in which the Services need to conduct a direct assessment of risk to a listed species. The GENEEC2 model has no provision for site-specific or region-specific inputs, such as soil characteristics, slopes, and meteorological data. Furthermore, with the development of simple-to-use implementations of PRZM/EXAMS for the farm pond and index reservoir (PRZM/EXAMS Express, Burns 2006), there seems to be little need for or practical value of GENEEC2. For Steps 2 and 3, EPA and the Services should be using region-specific or site-specific applications of PRZM/EXAMS or possibly more sophisticated watershed models.

As noted in Chapter 2 (see Table 2-1), the committee suggests a common approach that involves more refined and sophisticated modeling and analysis as one progresses from Step 1 to Step 3 in the ESA process. Given the current practices in exposure analysis and the need to estimate pesticide exposures and the associated spatial-temporal variations experienced by listed species and their habitats, the committee envisions the following stepwise approach to exposure modeling.

Exposure

- *Step 1 (EPA)*. Initial exposure modeling would answer the question, Do the areas where the pesticide will be used overlap spatially with the habitats of any listed species? The Services, which have extensive knowledge of the natural history of listed species, could help EPA to identify overlaps of areas where a pesticide might be used and the habitats of listed species. EPA's DANGER program would be useful in this step.
- *Step 2 (EPA)*. If area overlaps are identified in Step 1, EPA would confer with the Services to identify relevant environmental compartments (for example, pond vs stream), associated characteristics (for example, sandy vs silty soils), and critical times or seasons in which environmental exposure concentrations need to be estimated. With that knowledge, suitable model parameter values could be chosen and used. The goal of EPA's initial exposure modeling would be to identify the most important environmental compartments for exposure modeling (water, soil, air, or biota). Models—such as GENEEC2, SIP, and SPIR—would be useful in this step. If the models indicate that substantial amounts of pesticides move off the application site and into the surrounding ecosystems, more sophisticated fate and transport processes could be incorporated. At that point, the pesticide-fate model could be simplified to remove processes that are unimportant in the specific regions of the listed species and set up to estimate time-varying and space-varying pesticide concentrations in typical habitats (for example, 10-cm-deep shallow regions along streams vs 2-m-deep farm ponds) with associated uncertainties. The committee emphasizes that inputs should include statistical distributions of each parameter to enable probabilistic modeling of exposure scenarios. During Step 2, EPA could direct the terrestrial exposure modeling at specific size classes of taxonomic groups that represent the listed species of concern. On the basis of the modeling results, EPA could then make a decision about the need for formal consultation with the Services.
- *Step 3 (Services)*. During a formal consultation, the Services would further refine the exposure models to develop quantitative estimates of pesticide concentrations and their associated distributions for the particular listed species and their habitats. To that end, the models would use site-specific input values—for example, actual pesticide application rates, locally relevant geospatial data to characterize such quantities as wind speed and organic contents of soils, and time-sensitive life stages of listed species. The exposure analysis would be completed with propagated errors on exposure estimates.

Some issues associated with the exposure models or modeling practices need to be emphasized. First, pesticide-fate models are not always well tested with field data for specific pesticide applications at sites whose properties are knowable. Bird et al. (2002) tested AgDRIFT, and Loague and Green (1991) tested PRZM. However, a comprehensive treatment of the use of EXAMS with pesticides is largely lacking. Burns (2001) did list six studies involving field observations of diverse compounds that could be compared with EXAM modeling expectations, but none of the data involved pesticides applied in agricultural settings except the use of sulfonyl herbicides in rice fields. To evaluate and improve the accuracy of the exposure estimates, one could pursue a measurement campaign specifically coordinated with several pesticide field applications in a few case-specific examples during Step 3 exposure modeling. The exposure estimates should be compared with pesticide measurements in various environmental media, and modeling should be revised if measurements deviate substantially from selected statistical bounds, such as two standard deviations, of modeled estimates of environmental concentrations.

The committee notes that in evaluating models, general monitoring data and field studies need to be distinguished. General monitoring studies (see, for example, Gilliom et al. 2007) provide information on pesticide concentration in surface water or ground water on the basis of monitoring of specific locations at specific times. The monitoring reports, however, are not associated with specific applications of pesticides under well-described conditions, such as application rate, field characteristics, water characteristics, and meteorological conditions. General monitoring data cannot be used to estimate pesticide concentrations after a pesticide application or to evaluate the performance of fate and transport models.

Second, the model predictions can be only as accurate as the parameter estimates. If the relevant parameter values and their variances are poorly known, the model predictions will be uncertain and difficult to use in decision-making. That shows the need to identify the key processes and to ensure that the parameter values associated with the key processes are well known. The committee notes that although this is not typically done, exposure models can be used to identify the most important fate processes for a given pesticide application. For example, Sato and Schnoor (1991) used EXAMS to study the fate of dieldrin delivered by runoff to an Iowa reservoir. The pesticide's fate was dominated by flushing and bed-water exchange, so dieldrin exposures were sensitive to the depth of the mixed bed, and getting that parameter right was necessary to achieve accurate modeling. Similarly, Seiber et al. (1986) found that volatilization of 2-methyl-4-chlorophenoxyacetic acid from rice fields did not result chiefly from water-to-air exchanges but rather from transfers of salts dried on foliage to the air. Such key chemical fate processes, once identified, are almost never pursued in sufficient detail to allow substantial improvement in exposure modeling. Although studies by pesticide registrants might yield useful site-specific information, the empirical observations do not typically yield generalizable understandings of fate processes that can be readily used in new situations without introduction of further uncertainty.

Finally, the committee notes that the pesticide fate and transport models do not provide information on the watershed scale; they are intended only to predict pesticide concentrations in bodies of water at the edge of a field on which a pesticide was applied. Different hydrodynamic models are required to predict how pesticide loadings immediately below a field are propagated through a watershed or how inputs from multiple fields (or multiple applications) aggregate throughout a watershed. Watershed-scale models, such as the Soil and Water Assessment Tool (SWAT), have been developed to predict the effects of agronomic practices on water and sediment. SWAT operates on a daily time step and can perform simulations over a long time (30 years) by using physical landscape characteristics (including soil types and topography), data on land cover and land use, weather data, and physical-chemical properties of compounds to simulate processes that dictate routing of water and sediment. The primary routes for chemicals to enter water from a site of application in SWAT are surface runoff and infiltration of applied chemicals into groundwater that can reach surface waters through lateral flow and recharge. Thus, SWAT has an interface with PRZM/EXAMS or the Groundwater Loading Effects of Agricultural Management Systems (GLEAMS) (Leonard et al. 1989; Knisel and Davis 2000) model and can be used to predict chemical concentrations at particular points in a watershed over variable intervals.

GEOSPATIAL DATA FOR HABITAT DELINEATION AND EXPOSURE MODELING

Geospatial data are critical for exposure modeling and for describing species' habitats. The committee was asked to consider what constitutes authoritative geospatial data. The following sections discuss the delineation of habitat, describe the criteria for authoritative geospatial data, and provide several examples of various types of authoritative geospatial data.

Characterization and Delineation of Habitat

Habitat refers to the abiotic and biotic environmental attributes in an area that allow an organism to survive and reproduce (Hall et al. 1997). Habitat configuration, area, and quality—which vary over space and time—affect probabilities of persistence of populations and species. Because habitat by definition supports survival and reproduction, the term *suitable habitat* is redundant, and the term *unsuitable habitat* is contradictory. Habitat is species-specific, although a specific abiotic or biotic attribute might be a habitat component for multiple species; habitat is not synonymous with land cover, vegetation, or vegetation structure (Hall et al. 1997). Detailed explanations and discussions of the concept of habitat are included in Fretwell (1972), Morrison and Hall (2002), and Mitchell (2005). Characterization and delineation of species' habitats is necessary to estimate where and when a given pesticide and a given species might co-occur, to make spatially and temporally explicit calculations of pesticide exposure, and to specify the spatial structure of population models used in effects analyses.

The first step in delineating habitat is to compile data on species occurrence and, ideally, data on species' demography and environmental attributes that are associated with occurrence and measured in the field. Numerous publications have compared methods for identifying and statistically modeling associations between a species and its environment and have described the data requirements and the information content and potential applications of results (Scott et al. 2002; Elith et al. 2006; Franklin 2009; Royle et al. 2012). For example, resource-selection functions (Boyce et al. 2002; Manly et al. 2010) and occupancy models (MacKenzie et al. 2006) are among the diverse statistical methods that characterize habitat quality by relating data on the distribution or demography of a species to abiotic and biotic attributes of its environment. Regardless of method, the size of a species' range, and the specificity of its resource requirements, greater access to and reliability of geospatial data have made it easier to delineate and characterize habitat and habitat quality for a given species in space and time. The data also have improved the ability to model chemical fate and potential exposure of organisms. Horning et al. (2010) have presented a comprehensive, easily understood review of data sources and methods for application of remotely sensed data (data on an environmental feature that are not collected by physical contact with the feature) to ecological analyses.

Many caveats are associated with projections of habitat location and distributions of species. For example, most models of species distributions describe a statistical relationship between detections of an organism and elements of its habitat. The models tend to assume implicitly that species-environment relationships are stable—an assumption that might not be valid if habitat is currently unoccupied (Wiens et al. 2009) or if climate, land cover, or land use change (Araújo and Pearson 2005; Sinclair et al. 2010). Moreover, models of species distributions do not allow one to project species occurrence reliably in areas or periods in which environmental conditions are unsampled or otherwise unknown. Uncertainties increase if environmental data and species data were not collected in the same locations or during the same period. In addition, correlative models of species distributions do not account for phenotypic plasticity and adaptive evolution and therefore might overestimate reductions in range size in response to environmental change (Pearson and Dawson 2003; Skelly et al. 2007; Schwartz 2012).[2]

The level of uncertainty associated with a species' range and distribution and with delineation of its habitat is strongly affected by uncertainty in the data on species occurrence.[3] Ideally, data on occurrence are gathered over many years, in many locations that span the range of values of major environmental gradients, and with a sampling design that reflects the biology of the species.

---

[2]*Phenotypic plasticity* is defined as modifications of behavior, appearance, or physiology of individuals in response to environmental change, and *adaptive evolution* is defined as heritable genetic changes that affect individual phenotypes and increase probabilities of population or species persistence.

[3]*Range* is defined as the total extent of the area occupied by a species or the geographic limits within which it occurs, and *distribution* is defined as the areas in which a species is projected to occur on the basis of modeled associations with environmental attributes.

Such data might be collected during a sponsored research project but otherwise can be relatively rare. It often might be necessary to rely on such data sources as the North American Breeding Bird Survey, the Biodiversity Informatics Facility maintained by the Center for Biodiversity and Conservation of the American Museum of Natural History, and records on threatened or nonnative invasive species maintained by NatureServe (a nonprofit organization that represents an international network of data centers and state-level natural heritage programs). A number of uncertainties are common to atlases or databases of species occurrence (Franklin 2009), but they might represent the best data available in the absence of recent, standardized, or comprehensive field data on occurrence. Provided that uncertainties are estimated, statistical characterization and delineation of habitat is generally objective and quantitative and is more reliable than qualitative and subjective descriptions of habitat. In the event that decision-makers consider the uncertainties to be so high that new information must be collected, much guidance (Noon 1981; Buckland et al. 2001; MacKenzie et al. 2006; Willson and Gibbons 2009; Samways et al. 2010) is available about practical sampling methods for different taxonomic groups.

Criteria for Authoritative Geospatial Data and Metadata

The reliability of habitat delineations and ecological risk assessment is increased substantially by use of authoritative geospatial information and data (henceforth geospatial data) in which all parties have confidence and that all agree to use. Use of the same geospatial data by government agencies, nongovernment organizations, and private companies could facilitate joint fact-finding—a process through which diverse and sometimes adversarial parties collaborate to identify, define, and answer scientific questions that inform policy development (Karl et al. 2007).

Authoritative geospatial data should meet three criteria: they should be available from a widely recognized and respected source; they should be publicly available, whether freely or for purchase; and, for applications in the United States, they should be accompanied by metadata consistent with the standards of the National Spatial Data Infrastructure (NSDI). The criteria are applicable regardless of the scale of the data. Metadata document the fundamental attributes of data, such as who collected the data, when and where the data were collected, what variables were measured, how and in what units measurements were taken, and the coordinate system used to identify locations. Metadata allow one to understand a data source in sufficient detail to replicate the data collection and determine whether the data are applicable to a given analysis or decision-making process. The Federal Geographic Data Committee (FGDC 2012) and Dublin Core (DCMI 2012) maintain detailed technical and nontechnical explanations of metadata. Different federal agencies and research consortia have developed metadata standards that differ somewhat but remain consistent with the NSDI standards.

Standardized systems of data organization, storage, and retrieval facilitate compilation, discovery, accessibility, and assessment of the enormous amount of data on the arrangement and attributes of geospatial features and phenomena on Earth. The infrastructure of the NSDI includes the materials, technology, and people necessary to acquire, process, store, and distribute geospatial data to meet diverse needs (NRC 1993). Because the NSDI includes standards for geospatial data and specifications for metadata, all data in the archive are compatible regardless of source (FGDC 2007). The NSDI is administered by FGDC, an organization of federal geospatial professionals and constituents whose objective is to ensure that data can be efficiently shared among users and meet readily available standards.

Among the types of geospatial data most useful for delineating habitat and estimating exposure and effects of pesticides on listed species and their ecosystems are those on topography, hydrography, meteorology, solar radiation, soils, geology, and land cover. Although those data are not mutually exclusive, they generally are represented with different spatial-data layers. The sections that follow describe the various types of geospatial data and provide several examples of authoritative sources of them. In many cases, there are multiple authoritative sources of each type of data on different spatial and temporal scales. Although it would be ideal to be able to identify specific authoritative sources, no one authoritative data source will be best for all habitat delineations, exposure analyses, or other applications. However, accuracy assessments of authoritative data sources that are generally available might allow one to gauge which source is likely to be the most reliable for a particular objective. For example, the accuracy of a certain land-cover class might have higher priority than the accuracy of other classes, depending on the species or pesticide.

Topographic Data

Topographic metrics (such as slope, aspect, and elevation) often represent environmental features that are closely associated with species distributions (Osborne et al. 2001; Clevenger et al. 2002; Shriner et al. 2002) and that can affect chemical fate and transport. Diverse algorithms and modules within Geographic Information System software, such as ArcGIS modules (Environmental Systems Research Institute, Inc., Redlands, California), are available for modeling topography (Pelletier 2008; Horning et al. 2010).

Topographic features, such as heterogeneity of elevation in a given area or the boundaries of watersheds, can be derived from digital data on elevation. Sources of free elevation data include the National Elevation Dataset, the Shuttle Radar Topography Mission, and the Global Digital Elevation Map. Digital elevation models are available at resolutions of 30 m, 10 m, and, in some areas, 3 m.

Two free modules for ArcGIS—Topography Tools (ESRI 2010) and DEM Surface Tools (Jenness Enterprises 2011)—allow derivation of topographic data. For example, Topographic Position Index measures whether the elevation of a given pixel is greater or smaller than that of surrounding pixels. That information can be translated into values of slope that, in turn, can be used to model species-environment relationships (Dickson and Beier 2007). Topography also may be correlated with land uses, such as agriculture, residential development, and recreation.

Three-dimensional data acquired from light detection and ranging (lidar)—an optical remote sensing technology—afford many new ways to characterize vegetation, especially understory vegetation beneath tree canopies (Vierling et al. 2008), and to map the location and topography of flood plains and channels. ArcGIS modules, such as LIDAR Analyst (Overwatch Systems LTD 2009), enable processing and use of lidar data for developing accurate models of land-surface features at spatial resolutions relevant to many modeling applications (for example, less than one to tens of meters). Models of elevation and above-ground measures of vegetation structure derived from lidar data are increasingly used to model species' habitats and distributions (Bradbury et al. 2005; Martinuzzi et al. 2009). The US Geological Survey (USGS) Center for LIDAR Information Coordination and Knowledge is intended to improve access to lidar data and coordination among and education of its users (USGS 2012a).

Hydrographic Data

Watershed features are relevant to habitat delineation of terrestrial and aquatic species and to assessment of potential pesticide exposure of these species. For example, there might be fewer natural barriers to movements of species and toxicants along river banks and within watersheds than between watersheds. A national system of hydrologic unit codes (HUCs) divides the United States into six nested sets of watersheds; that is, large watersheds are progressively divided into smaller watersheds (Seaber et al. 1987). At its coarsest resolution, the HUC system delineates 21 regions that are large watersheds (such as the Rio Grande) or logical groups of similar drainages (such as the Pacific Northwest, California). Each region is labeled with a name and a two-digit number; for example, the Columbia River Basin is numbered 17. As HUCs are subdivided, each subdivision is labeled with a name and an additional two digits; for example, the combined Kootenai, Pend Oreille, and Spokane river basins correspond to number 1701, and the Kootenai River Basin is numbered 170101. The smallest hydrologic units, subwatersheds, have 12-digit labels (Table 3-1). Hydrologic units span nearly 5 orders of magnitude in size, from about 100 km^2 (40 mi^2) for subwatersheds to about 460,000 km^2 (178,000 mi^2) for regions. In some parts of the country, watersheds have been delineated at resolutions as fine as 16-digit HUCs (NRCS 2012a).

The standardized watershed boundaries of the HUC system provide a common geographic context for all users. The boundaries are available from USGS on paper maps (USGS 2010a) or in digital form (USGS 2012b). The metadata for the digital data and a description of the philosophical foundation of the system also are available at no cost (USGS/USDA/NRCS 2011). Overlaying hydrographic and topographic data sometimes reveals inaccuracies in the geographic locations of small streams, but these inaccuracies typically can be resolved with aerial photographs or field validation.

Substantial amounts of data associated with six-digit HUCs are available on-line from EPA (2012e). The data are diverse and include social variables, such as human demography, and ecological variables, such as water quality. Data are provided in formats and with documentation that do not require substantial technical expertise to understand or apply.

Some states maintain an accounting system for water resources separate from the federal HUC system. For example, the Washington Department of Ecology defines water resource inventory areas (WA Department of Ecology 2012). The boundaries of the inventory areas are not identical with those defined by the HUC system, but the inventory areas have some historical precedent in the state. A map of the inventory areas also serves as a graphical user interface to access many types of data associated with the biology and management of listed species (WA Department of Ecology 2012).

After defining a watershed, one can classify the relative size and location of its constituent streams (Ritter et al. 2011). In this classification system, the smallest tributaries are assigned the order of 1 and referred to as first-order streams. When two first-order streams join, they continue as a single stream of the second order. When two second-order streams join, they form a single third-order stream, and so forth. Low-order streams (small numbers) are always in headwater regions, whereas high-order streams are main rivers. Stream ordering is not highly amenable to quantitative analysis because its application depends on the resolution at which an observer perceives the landscape. Small maps showing large areas, for example, might omit first-order streams that are apparent in field observations.

TABLE 3-1 Nested Hierarchy of Hydrologic Units

Number Digits in HUC[a]	Hydrologic Unit Name	Number of Units	Average Size of Unit in km^2 (mi^2)
2	Region	21	459,878 (177,560)
4	Subregion	222	43,512 (16,800)
6	Accounting unit[b]	352	27,454 (10,600)
8	Cataloging unit[b]	2,150	1,813 (700)
10	Watershed[c]	~20,000	588 (227)
12	Subwatershed[c]	~100,000	104 (40)

[a]Hydrologic unit code.
[b]Numbers of units and boundaries revised from Seaber et al. (1987) by later users.
[c]Mapping not yet complete.
Source: Seaber et al. 1987, later revised and reported by USGS 2011.

Exposure

Meteorological Data

Variation in weather at relatively small spatial resolutions (such as kilometers to tens of kilometers) and temporal resolutions (such as days to a few years) can affect the distributions and population dynamics of organisms and their resources. Chemical fate and transport also are affected by meteorological variables, such as temperature, precipitation, and wind speed and direction. Accordingly, those variables will affect chemical-fate model parameters, such as probability of runoff and loads of suspended solids.

Meteorological data in the United State are archived and made freely available by national and regional centers maintained by NOAA. The National Climatic Data Center has complete data on the 122 primary National Weather Service (NWS) reporting stations in the United States.[4] Gridded climatic data are also available for a variety of cell sizes (ESRL 2012). The six regional climate data centers provide the same data as the national center and observations or estimates of regional relevance (NCDC 2012). The meteorological data available through the national and regional sources are authoritative in that they were collected by the NWS or its partners, have been screened and checked by experts, are accompanied by complete metadata, and are publicly available. The data are available in tabular format and in a spatial format that meets the NSDI standards.

Solar Wavelength and Radiation Data

Solar radiation at wavelengths of about 290–600 nm affects rates of photochemical excitation and transformations and therefore chemical decomposition of pesticides. Data on solar radiation are used to calculate insolation—the amount of solar radiation that reaches a given location on Earth's surface—which affects cyclic or seasonal phenomena, such as migration; rates of growth and development; and the distributions of species in space and time.

Daily data on the distribution of solar wavelengths from the ultraviolet to the near infrared are available from the National Aeronautics and Space Administration Earth Observing System's Solar Radiation and Climate Experiment.[5] However, the measurements of incoming solar radiation are taken at the top of the atmosphere rather than at Earth's surface. The distribution of wavelengths received at Earth's surface are a function of latitude, day of the year, time of day, slope and aspect of the surface, cloud cover, concentrations of aerosols in the atmosphere, and horizon obstruction (Rich et al. 1994). Therefore, without surface measurements, calculation of direct photolysis rates and half-lives of chemicals in water and on soil surfaces requires estimation of numerous atmospheric parameters and use of those parameters and spatial coordinates, time of

---

[4] See www.ncdc.noaa.gov/oa/ncdc.html.
[5] See http://lasp.colorado.edu/sorce/.

year, and time of day in a computer model, such as GCSOLAR (EPA 2012f) or SMARTS (NREL 2012). It might not be feasible to implement such models for all pesticides. Thus, existing geospatial data might not be sufficient to model some aspects of chemical fate. However, applying a model, such as SMARTS, to a given region and period (for example, the Pacific Northwest in spring) might allow one to determine the variability of the light intensity at the relevant wavelengths—those at which a given compound has high absorptivity. If exposure analysis suggests that photolysis is highly relevant to chemical fate, characterizing that variability would probably be valuable.

Insolation is calculated on the basis of Julian day and the coordinates and slope of the surface. ArcTools (ESRI, Redlands, California) also offers multiple tools for computing insolation for polygons or points. The Solar Radiation Graphics tool in ArcTools allows one to visualize the visible sky, the sun's position in the sky over time, and the sectors of the sky that affect the amount of incoming solar radiation, all of which are incorporated into calculations. The Photovoltaic Education Network provides an on-line calculator,[6] which is authoritative in that it is a product of an organization that provides training for the solar engineering industry, its calculations are freely available, and the metadata provided on the site explain how the calculations are derived.

Soils Data

Soil type is associated with habitat quality for wild plants and agricultural crops and for animals that communicate by pheromones and other chemicals. Chemical fate might be associated with soil infiltration and runoff, and soil pH and anion-cation exchange capacities of soils are useful parameters for modeling sorption.

In the United States, the authoritative source for soils data is the Natural Resources Conservation Service (NRCS; formerly the US Soil Conservation Service) of the US Department of Agriculture (USDA). Since the 1930s, NRCS has mapped almost all the soils in agricultural areas in the country. Soils data are available as digitized maps accompanied by narrative descriptions and some numerical data about the soils (NRCS 2012b). Because almost all the original soil surveys were conducted at the county level, the data are organized by county. The base maps typically are aerial photographs on which polygons that represent different soil types are superimposed. Each soil type has a distinct identifier. Soil classification is conducted by interpreting aerial photographs and field surveys. The resolution of the resulting maps is sufficient to identify the soil type in individual fields.

The narrative for each county's soils contains quantitative information about particle sizes, basic soil chemistry, organic content, and hydrologic attributes. The narratives also describe soil horizons, which are multiple layers of soil

---

[6]See pvcdrom.pveducation.org/SUNLIGHT/MODTILT.HTM.

Exposure 63

below the surface. Field measures of soil properties might be necessary for some model applications, but the NRCS soil surveys typically are adequate for models that require values of basic soil attributes. The NRCS soils data are authoritative in that they are products of USDA and the work of experts in soil science. The data are freely available and meet NSDI standards, and metadata are complete.

Geological Data

Geology strongly influences the chemistry of surface materials and shallow groundwaters that interact with pesticides. Authoritative geospatial data on geology in the United States are provided by USGS via its Mineral Resources Online Spatial Data (USGS 2012c) and to a lesser extent by the offices of state geologists or state geological surveys. For example, Washington state provides geospatial geological data on the distribution of rock units, including rock types and the geological age of each unit (Dragovich et al. 2002). Further information about the physical characteristics of each rock unit is published by USGS or its state counterparts.

The geology of the entire United States has been mapped on some scale. In most cases, geological maps are available at the resolution of counties; in some areas, the map scales are as fine as 1:24,000. USGS maintains a Web site with an interactive map of the United States that is linked to geological data on each state in a variety of formats (USGS 2012c). The site also links to complete metadata for each state, publications that describe the methods used to generate geological maps, and narrative descriptions of the physical and chemical properties of surface and subsurface rocks. Geospatial data on geology were collected to support numerous activities, such as mineral exploration, detection of faults, oil and gas exploration, and designation of national parks. As a result, there is considerable variation in the supporting documentation and narrative descriptions of the maps.

Nationwide geological data reflect more than a century of detailed mapping and analysis by expert geologists. The metadata are extensive, data and narratives are freely available, and the maps adhere to the standards of the NSDI.

Land-Cover Data

Land cover encompasses both natural features—such as native vegetation, rock formations, and bodies of water—and features produced by human activity, such as agricultural fields and urban areas. Types and quantities of pesticides applied sometimes can be inferred on the basis of the distribution of crop types. Delineating habitat for some species or assessing particular exposures might not be possible on the basis of existing classifications of land cover. Depending on the species biology or the pesticide characteristics, it might be necessary to develop a new classification of regional land cover on the basis of satellite images, aerial photographs, and field validation. For example, although the boundaries of agricultural fields might be stable over many years, crop types might vary among and within years. When a time series of land-cover data is available, it might be possible to develop a spatially explicit probability distribution of changes in all or a subset of land-cover types. Features of agricultural land that might be attributes of habitat for some species, such as small groups of trees or streams, typically are not included in publicly available crop data. However, in many cases, it is sufficient to derive land-cover data from another source. Whether a new classification is necessary depends on the target location, species, and pesticides; the focus of the assessment, which will determine the relevant cover types and spatial and temporal scale of the data needed; and the necessary level of data accuracy.

USGS provides numerous sets of land-cover data that cover the conterminous United States and smaller areas, such as selected states or ecosystems (USGS 2010b). More detailed data are available on some land-cover types, such as wetlands and forests. Among the most commonly used sets of land-cover data derived from Landsat images at 30-m resolution are the National Land Cover Dataset and those available from the National Gap Analysis Program (Scott et al. 1993, 2002) and the Landscape Fire and Resource Management Planning Tools (LANDFIRE) project. Regional programs, such as the Southwest Regional Gap Analysis Project, offer seamless maps—which do not change abruptly at state boundaries—of land cover across multiple states with climate and species composition that are distinct from elsewhere in the nation. Both national and regional gap-analysis programs provide projections of the current ranges and distributions of multiple taxonomic groups. For example, the national program includes range data on 1,376 species of amphibians, birds, mammals, and reptiles and distribution data on 810 species (USGS 2011).

The National Agricultural Statistics Service provides spatial data and metadata on the distribution of 133 classes of land cover, including major crop types, across the country (NASS 2010). Principal sources of raw data for its classification are the Resourcesat-1 Advanced Wide Field Sensor and Landsat Thematic Mapper. National data are available on each year since 2008; data on 2011 are at 30-m resolution. Annual data on some states extend back to 1997. The Web-based application CropScape (Han et al. 2012) is a user-friendly interface with the data.

State and local sources of spatial data on agricultural land use vary. For example, since 1984, the California Farmland Mapping and Monitoring Program has tracked the distribution of agricultural land and urban development (CA Department of Conservation 2007a). The program releases spatial data every 2 years with a minimum mapping unit of 4 hectares. Data sources include aerial photographs, public comments, and field surveys. Among the land-use classes are grazing land, urbanized land, four types of farmland, and five types of rural land (CA Department of Conservation 2007b).

UNCERTAINTIES IN EXPOSURE MODELING AND PARAMETER INPUTS

The chemical-fate models with such diverse information as geospatial data can be used to obtain an EEC. Many uncertainties are associated with that estimation, and this section explores some of the most important ones and suggests methods for addressing them.

Pesticides and Mixtures

The first requirement for successful exposure modeling involves identification of the specific substances that are to be introduced into the environmental setting. Those data are needed not only to evaluate exposures to individual components but to assess prospective interactions of the components. To have an informed discussion on pesticide exposure, three types of mixtures need to be distinguished.

Pesticide formulations. Typically, a pesticide manufacturer or supplier mixes one or more active ingredients—the chemicals that are responsible for a pesticide's biological effects—and other chemicals. The mixture is what is often referred to simply as the pesticide or the pesticide formulation. The committee notes that virtually no chemical is synthesized as a pure compound, so impurities occur in the synthesis of the pesticide active ingredients. Although manufacturing processes try to reduce the number and concentrations of impurities, technical-grade active ingredients that are used to make the pesticide formulations will contain the active ingredients and some impurities.

Tank mixtures. In most pesticide applications, pesticide formulations are added to a tank or other container with adjuvants (see below). The term *tank mixture* refers to the material in a tank or container at the time that the material is applied to a treatment area, such as an agricultural field. Exposure issues associated with pesticide formulations and tank mixtures share a property that greatly simplifies exposure analysis—the materials are applied at the same time to a defined location. More important, the identity and concentration of the constituents are known.

Environmental mixtures. This term is used to designate all contaminants that are in the environmental media of concern, such as water in the case of salmonids. Environmental mixtures are the results of previous applications of tank mixtures—sometimes many tank mixtures applied at different times to different areas in a watershed or other locale of concern. In addition, environmental mixtures include other environmental contaminants not related to pesticide applications in the media of concern. Because environmental mixtures are the results of many sources of contamination, estimating the components in environmental mixtures quantitatively is far more difficult than estimating exposures associated with the application of a single tank mixture.

For pesticide risk assessments, EPA typically focuses its assessments on the active ingredients, whereas the Services contend that all the other chemicals or whole products need to be considered. The following sections describe in further detail the types of mixtures potentially involved, their components, and difficulties encountered in incorporating them into an exposure analysis.

Pesticide Formulations and Tank Mixtures

Pesticide formulations typically contain chemicals other than the active ingredients that often do not have a direct effect on the target species. The term *inert* is used to designate a chemical that is not classified as an active ingredient. Some inerts can be toxic, and EPA has proposed the term *other ingredients* rather than *inerts* (EPA 2012g). Nonetheless, *inert* is engrained as a term in the pesticide literature and is commonly used—for example, the EPA Inert Ingredient Assessment Branch, which was established in 2005. For brevity, the following discussion uses the term *inert* but recognizes that inerts might be biologically active and potentially hazardous.

The term *adjuvant* is closely related. Adjuvants differ from inerts only in that adjuvants are added to a tank mixture in the field at the time that the pesticide is applied rather than when it is formulated. Tank-mixture adjuvants—such as surfactants, compatibility agents, antifoaming agents, spray colorants (dyes), and drift-control agents—are added to a tank mixture to aid or modify the action of a pesticide or the physical characteristics of the mixture (Ferrell et al. 2008).

Inerts and adjuvants are an extremely broad array of chemicals, including carriers, stabilizers, sticking agents, and other materials added to facilitate handling or application. Mixtures of different pesticide formulations or pesticide formulations in combination with various adjuvants are typically applied to save time and labor and to reduce equipment and application costs. Such a mixture might also control a variety of pests or enhance the control of one or a few pests.

EPA is responsible for the regulation of inerts and adjuvants in pesticide formulations. EPA (52 Fed. Reg. 13305 [1987]) developed four classes (lists) of inerts on the basis of the available toxicity information: toxic (List 1), potentially toxic (List 2), unclassifiable (List 3), and nontoxic (List 4). List 4 was subdivided into two categories: List 4A contained inerts on which there was sufficient information to warrant a minimal concern, and List 4B contained inerts the use patterns of which and toxicity data on which indicated that their use as inerts was not likely to pose a risk. Although EPA no longer actively maintains the lists, references to that classification system are in the older literature; moreover, EPA documents, such as the current Label Review Manual (EPA 2010), still refer to the lists of inerts.

After a review of the toxicity data that supported food tolerances for pesticide inerts, EPA (71 Fed. Reg. 45415[2006]) revoked food tolerances for over 100 inerts; that is, these inerts can no longer be used in pesticides that are applied to food commodities. Thus, no List 1 inerts are now allowed in food-use pesticide formulations. Only five of the original 50 List 1 inerts—di-*n*-octyl adipate; ethylene glycol monoethyl ether; 1,2-benzenedicarboxylic acid, bis(2-ethylhexyl) ester; 1,4-benzenediol; and nonylphenol—are now permitted in nonfood-use pesticide formulations (EPA 2011). In 2011, EPA released a searchable on-line database of inerts that are approved for use in pesticide formulations (EPA 2012h). The database includes three sets of inert ingredients: those approved for food and nonfood use, for nonfood use only, and for fragrance use.

Some inerts used in pesticide formulations are complex mixtures, for example, petroleum-based solvents and tallow-based surfactants. Petroleum hydrocarbon solvents can contain many individual compounds, and the compositions of the solvents vary substantially, depending on the distillation process and on the sources and types of the crude oil used to derive the petroleum distillates (ATSDR 1999). Similarly, surfactants based on tallow (animal fat) are highly complex mixtures whose compositions vary on the basis of the source of the animal fat and the manufacturing processes used to render the animal fat and process the tallow (Kosswig 1994; Brausch and Smith 2007; Katagi 2008).

In some cases, applications of multiple pesticide formulations and tank mixtures might not present difficulties in the exposure analysis beyond those associated with applications of a single compound. If components of a tank mixture or formulation do not substantially affect the fate and transport of other components, the exposure analysis methods used for single chemicals can be applied to tank mixtures. In cases in which additives, such as surfactants, could affect the fate and transport of active ingredients in a formulation, uncertainties in exposure analysis could be substantial unless the effect of additives can be quantified in exposure modeling. Many inerts are designed to affect the behavior of an active ingredient after application. For example, surfactants or penetrating agents are often used in applications of herbicides. Surfactants and penetrating agents might have little or no phytotoxicity at the concentrations used in most herbicide applications, but their ability to enhance absorption can enhance the efficacy of herbicides (Denis and Delrot 1997; Liu 2004; Tu and Randall 2005). Surfactants can also alter the persistence and mobility of active ingredients in soil and water (Katagi 2008); similarly, microencapsulation can retard transport in soil. Prolonging residence time can enhance the efficacy of pesticide active ingredients in soil (Beestman 1996).

The environmental-fate properties of inerts often differ from the corresponding properties of a pesticide's active ingredients. Consequently, inerts and active ingredients partition differentially in the environment (water, sediment, soil, and air) and do so at differing rates. Individual constituents in complex inerts also have different environmental-fate properties, so components of the inerts also partition at different rates and to different extents. Little information is available on the environment fate and differential partitioning of complex inerts. Although a relatively standard set of tests are required on the environmental fate of most active ingredients, testing requirements are less stringent for inerts.

Environmental Mixtures

Unless exposure occurs only at or near the point of pesticide application, species are more likely to be exposed to environmental mixtures than to a single pesticide formulation or tank mixture. Environmental mixtures are formed when a tank mixture—active ingredients, inerts, and adjuvants—combines with other chemicals that are already present in the environment from other sources, such as other pesticides from previous applications and pharmaceutical, consumer, and personal-care products in municipal effluent.

As a formulation or tank mixture moves away from the initial point of application, its components often do so at different rates and exhibit differential partitioning into various environmental media (surface soil, surface water, sediment, and air) and undergo transformations—for example, fipronil to its more toxic and persistent degradates (Lin et al. 2009)—at different rates. The chemical components become diluted in environmental media that already contain other chemicals, including pesticides. For example, in Oregon's Willamette River Basin, only 3.6% of surface-water samples collected during 1994-2010 contained only a single detected chemical; over 50 pesticide mixtures of two to six pesticides each were found in the remaining samples (Hope 2012). Nationally, more than 6,000 unique mixtures of five pesticides were detected in agricultural streams (Gilliom et al. 2007). The data from Gilliom et al. (2007) are cited in the BiOps (NMFS 2008, 2009, 2010) as a basis for documenting that exposures to environmental mixtures will occur. The monitoring data from Gilliom et al. (2007), however, are not associated with specific applications of pesticides.

Approaches to estimating exposures to environmental mixtures are at least conceptually similar to those associated with pesticide formulations or tank mixtures. If the exposure factors are known—that is, the pesticide and environmental components, their concentrations, and their locations at a specific time—exposure-analysis methods can be used to assess exposures to the environmental mixture. In practice, however, quantitative estimates of exposures to environmental mixtures are seldom feasible owing to the dynamic state of the environmental mixtures and the varying compositions of the mixtures over space and time. In any given location or watershed, the amounts of pesticide active ingredients, inerts, and adjuvants in environmental media will be highly variable and depend on pesticide use and other sources of environmental contamination.

As noted by the Services, the EPA pesticide risk assessments do not directly or explicitly incorporate information on exposures to environmental mixtures. The Services commonly address environmental mixtures in the assessment of the *baseline* (the state of a population excluding exposure to the pesticide under consideration), but these considerations are largely qualitative rather than quantitative. Although all the BiOps discuss available modeled estimates and monitoring data on multiple pesticides that might occur as environmental mixtures (see, for example, NMFS 2011, Table 107), this information is not used

Exposure 69

quantitatively to modify risk assessments that focus on exposure to one or more active ingredients. NMFS (2011, p. 442) notes that "given the complexity and scale of this action, we are unable to accurately define exposure distributions for the chemical stressors." Essentially the same language is included elsewhere (NMFS 2008, p. 259; 2009, p. 309; 2010, pp. 449-450).

The qualitative discussions of exposures to environmental mixtures in BiOps by the Services and the focus of EPA on single chemicals are not fundamentally different. EPA's basic agreement with the position taken by NMFS is clearly illustrated in its response to questions posed by the committee (EPA 2012i, p. 5), which included the following:

> The highly variable nature of the background exposure to other chemical stressors represents a significant impediment to combined effects analysis. Much of the empirical data for multiple chemical stressor evaluation involves small suites of chemicals, in discrete concentration combinations that are not highly representative of in-field conditions across complex landscapes at the national scale of pesticide use that EPA must assess. In addition, predicting the frequency and pattern of environmental mixtures at the temporal scales used in acute and chronic risk assessment (hours to a few weeks) is beyond the capabilities of the best available nationwide data sets that look at combined chemical analysis.

The statements by EPA and NMFS above are functionally identical with respect to the qualitative rather than quantitative treatment of environmental mixtures.

The Services (see, for example, NMFS 2008, 2009, 2010, 2011) and other analysts (for example, Hoogeweg et al. 2011) often discuss or assess the potential co-occurrence of various pesticides (that is, pesticide active ingredients) with populations of listed species, but quantitative analyses of the co-occurrence of multiple pesticides have not been encountered in EPA assessments. As discussed in Gilliom et al. (2007, p.81), an analysis of the co-occurrence of pesticides might be useful in identifying environmental mixtures that have the greatest probability of adversely affecting listed species, and these investigators provide a preliminary assessment of the most commonly occurring mixtures of two to seven pesticides (Belden et al. 2007). More detailed analyses of the frequency of the co-occurrence of pesticides have been used in human health risk assessments (e.g., Stackelberg et al. 2009; Tornero-Velez et al. 2012). The preliminary analyses by Belden et al. (2007) on pesticides associated with corn and soybean production suggest that factoring the occurrence of environmental mixtures into assessments will increase the risk estimates but not substantially (by a factor of about 2). Although some BiOps (NMFS 2008, 2009, 2010) cite the analysis by Belden et al. (2007), they do not attempt to model exposures to multiple pesticides in a single watershed.

Pesticide Application Rates

Pesticide application rate is another important source of uncertainty. Despite a label's explicit application specifications, such as 1 lb of material per acre for corn fields, users commonly apply lower quantities according to the severity of their weed or pest infestation. However, Steps 1 and 2 of the ESA process (Figure 2-1) should ensure that no potentially unsafe pesticide applications are ignored. Accordingly, an exposure modeler can only assume that a given pesticide is applied at the maximum allowable rate. Furthermore, in Step 3 of the process (Figure 2-1), the Services cannot reasonably be expected to use information that suggests that substantially lower application rates are used unless supporting data are available. Such data must include statistical descriptions of the spatially and temporally distributed application rates. Moreover, some measures would have to be taken to ensure that a use pattern could not dramatically increase in any particular season or locale (for example, because of crop shifts). Only then could exposure modelers use such knowledge to obtain EECs with associated uncertainties. For now, pesticide use is probably an inaccurate input for exposure analysis; registration and labeling are not well suited for solving this exposure-analysis bias.

Other Fate-Modeling Parameters

Even if release rates per unit area of all the pesticide components were well quantified, other phenomena add uncertainty to estimates of exposure of various environmental surfaces, such as plant surfaces, soil surfaces, and surface water. For example, AgDRIFT includes numerous and diverse parameters (see Box 3-1). The certainty with which each relevant parameter for a particular pesticide application is known will influence the certainty of estimated chemical loadings on foliage, soil surfaces, and even neighboring surface waters. Bird et al. (2002) compared field data with AgDRIFT model evaluations for "161 separate trials of typical agriculture aerial applications under a wide range of application and meteorological conditions." The comparisons all relied on case-specific meteorological data (wind, temperature, and humidity) and application data, such as observed aircraft heights and nozzle equipment. With such inputs, the investigators concluded that the "model tended to overpredict deposition rates relative to the field data for far-field distances, particularly under evaporative conditions" by about a factor of 3. However, the AgDRIFT estimates were in good agreement (to within less than a factor of 2) with "field results for estimating near-field buffer zones needed to manage human, crop, livestock, and ecological exposure." Overall, aggregating the data for the various application methods resulted in ratios of model predictions to field measures of $10^{-0.03 \pm 0.5}$ at 23 m and $10^{0.10 \pm 0.9}$ at 305 m, given as $10^{\text{mean} \pm 1\text{SD}}$, where SD is standard deviation. Therefore, despite simplifying assumptions and the variability of some of

Exposure

the input parameters, one might conclude that the model itself operates fairly accurately. Bird et al. (2002) concluded that "the model appears satisfactory for regulatory evaluations." However, greater uncertainty in the output of the model will arise when it is applied as a general screening tool and case-specific input parameters, such as wind speeds and mode of application, are not known. That situation is true for other complex models, such as PRZM/EXAMS. One option for improving the situation is to use relevant geospatial data for estimating relevant fate-modeling parameters and their variability.

In addition to inaccuracies and imprecisions in initial pesticide loadings on soil, parameters used in chemical-fate models, such as PRZM and EXAMS, have associated uncertainties, particularly because pesticides often contain or are applied with other chemicals that can affect some fate processes. Data sources for assigning values to parameters range from empirical observations reported by pesticide registrants to information extracted from peer-reviewed journal publications that sought to elucidate underlying process mechanisms. As illustrations, consider two processes that are typically important in chemical-fate modeling: sorption and biodegradation. Both have been studied intensively for decades.

BOX 3-1 AgDRIFT Inputs

Aircraft information
    Aircraft type (fixed-wing, biplane, helicopter)
    Aircraft semispan or rotor radius
    Spraying speed
    Rotor-blade RPM (helicopter)
    Aircraft weight

Propeller Information
    Aircraft drag coefficient
    Aircraft platform area
    Engine efficiency
    Propeller RPM
    Propeller-blade radius
    Propeller location

Nozzle information
    Number of nozzles
    Nozzle type
    Nozzle locations

Drop size distribution

Spray-material information
    Tank-mix specific gravity
    Tank-mix flow rate
    Tank-mix nonvolatile fraction
    Tank-mix active fraction
    Evaporation rate

Meteorological information
    Wind speed
    Height of wind-speed measurement
    Surface roughness
    Wind direction
    Wet-bulb temperature depression (temperature and relative humidity)

Other information
    Spraying height
    Number of swaths
    Swatch width
    Swath displacement

Source: Teske et al. 2002.

Sorption phenomena are generally well understood, and sorption coefficients can be estimated relatively well in many cases. Assuming application of the correct sorption model (see below), sorption inputs in pesticide-fate models probably have only a moderate uncertainty (a factor of 3). For example, sorption coefficients (K_d values) can typically be estimated for nonionic organic compounds by using the product, $f_{oc}K_{oc}$, in which f_{oc} is the organic carbon content of the soil or sediment (kg_{oc}/kg_{solid}) and K_{oc} is the organic carbon-normalized sorption coefficient (mol kg^{-1}_{oc} / mol L^{-1}_{water}). In a review of the literature, Gerstl (1990) found that K_{oc} values for atrazine are log-normally distributed and vary only by about a factor of 2 (±1 SD) for 217 reported measurements of atrazine (Figure 3-1; $K_{oc} = 10^{2.1 \pm 0.3}$ given as $10^{mean \pm 1\ SD}$). That result is similar to the factor of 2.5 found by Seth et al. (1999) and suggested in the EXAMS user manual. It is also consistent with observations reported by Novak et al. (1997) for atrazine sorption in a single field site (Figure 3-2). Consequently, the sorption coefficient, K_d (L/kg_{solid}) for a specific soil or sediment, calculated by using the f_{oc} of that solid, can be known almost as precisely as the pesticide's K_{oc} values because site-specific f_{oc} measures can be made with great precision. However, if model calculations use a generic value for f_{oc} or even a value based on regional soil mapping (see section "Geospatial Data for Habitat Delineation and Exposure Modeling" above), one can readily anticipate deriving another factor of 2 from "real-world" variability around the f_{oc} term used to make the K_d estimate.

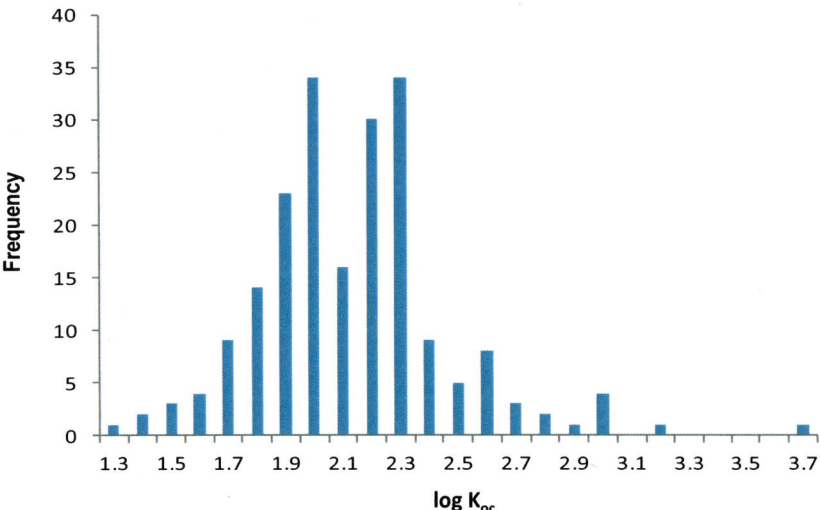

FIGURE 3-1 Organic-carbon normalized sorption coefficients, K_{oc}, values for atrazine plotted on a logarithmic scale. Source: Gerstl 1990. Reprinted with permission; copyright 1990, *Journal of Contaminant Hydrology*.

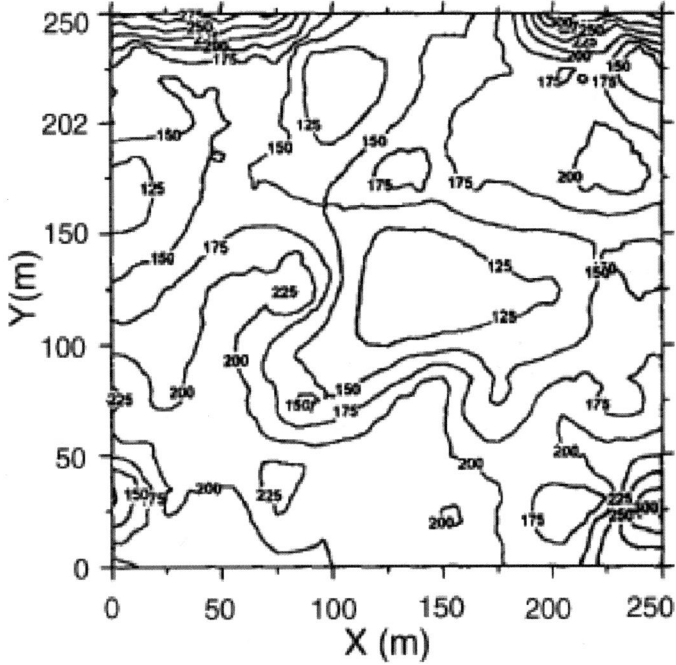

FIGURE 3-2 Distribution of K_{oc} values for atrazine in a 6.25 hectare field, showing a range of about a factor of 2. Source: Novak et al. 1997. Reprinted with permission; copyright 1997, *Journal of Environmental Quality*.

Perhaps more important, larger inaccuracies in predicting the amount of chemical sorbed to soil or sediment particles will result if the model used to describe the sorption process is inaccurate. For example, one cannot expect an accurate result if one uses a sorption model designed for nonionic pesticides ($K_d = f_{oc}K_{oc}$) when modeling ionic compounds. Some modelers made that mistake with the herbicide 2,4-D, which typically exists in water as a negatively charged species. A modeler should expect its sorption to involve anion exchange, as has been shown by Hyun and Lee (2005). A second case of an inappropriate use of the $f_{oc}K_{oc}$ model involves situations in which black carbon sorbents play an important role in addition to the rest of the organic carbon. Yang and Sheng (2003) have provided evidence of such sorption to black carbon in the case of diuron applied to a field with burned wheat and rice residues. Thus, although cases that accurately use the $f_{oc}K_{oc}$ model probably reflect modest levels of uncertainty (1 SD, reflecting a factor of 2-4), pesticide-fate modelers should recognize both chemical-specific properties and site-specific conditions that can cause their estimates of sorption to be quite inaccurate—not merely imprecise but biased by a factor of 10—when such a sorption model is inappropriately used (Accardi-Dey and Gschwend 2002).

Even the best estimates of biodegradation coefficients are generally much more uncertain than sorption estimates. When Laskowski (1995) reviewed literature on biodegradation rates in soils, he found that for many chemicals few soils were tested (Table 3-2). However, when substantial data were available, biodegradation rates varied widely, often by more than a factor of 10 (Table 3-2). Likewise, Paris et al. (1981) found that the rates of biohydrolysis of the butoxyethyl ester of 2,4-D varied by up to a factor of 25 in 33 water samples tested even when efforts were made to account for sample-to-sample variations in microbial population densities. Finally, in some cases, such as nitrilotriacetate, Tiedje and Mason (1974) observed significant lag periods (4-6 days) in three of 11 soils tested. Thus, inaccuracies will result if a simple first-order removal-rate law with a single-value rate coefficient is used for periods that are shorter than or comparable with such lag periods.

TABLE 3-2 Variability of Pesticide Degradation Rates in Soils

Pesticide	No. Soils Tested	Ratio of Highest to Lowest Degradation Rate Observed	Reference
Nitrilotriacetate	11	80	Tiedje and Mason 1974
Crotoxyphos	3	36	Konrad and Chesters 1969
Carbofuran	4	25	Getzin 1973
Glyphosate	4	19	Rueppel et al. 1977
Flumetsulam	21	10	Lehmann et al. 1992
Chlorimuron ethyl	19	8	L.M. Kennard and D.A. Laskowski, DowElanco, unpublished material, 1992
Thionazin	4	7	Getzin and Rosefield 1966
Nitrapyrin	10	6	Laskowski and Regoli 1972
Imazaquin	3	5	Basham and Lavy 1987
Chlorsulfuron	8	4	Walker et al. 1989
Methidathion	4	3	Getzin 1970
Aldicarb	2	2	Richey et al. 1977
Diazinon	4	2	Getzin and Rosefield 1966
Linuron	4	2	Lode 1967
Methomyl	2	2	Harvey and Pease 1973
Propyzamide	5	2	Walker 1976

Source: Adapted from Laskowski et al. 1982; Laskowski 1995.

Exposure 75

In the few cases in which sufficient data are available, it appears that biodegradation rates are log-normally distributed (Figure 3-3). For example, the (pseudo-) first-order rate coefficients reported by Lehmann et al. (1992) for flumetsulam appear to have a log-normal distribution (Figure 3-3, N = 21). Paris et al. (1981) also found that the microbial population-normalized rate coefficients appeared to be log-normally distributed with a k_{bio} value of about $10^{0.7 \pm 0.3}$ L/colony forming units (cfu) per hour ($10^{mean \pm 1SD}$, N = 33) for biohydrolysis of the butoxyethyl ester of 2,4-D. In both cases, the data suggest uncertainty of about a factor of 2 (about ±0.3 log units around the mean). To summarize, pesticide-exposure analysis should be pursued by using enough biotransformation information to establish whether the rates are normally or log-normally distributed, and then the data should be analyzed to obtain the mean rate coefficient and its variance for use in fate modeling.

It is also clear that rates of biodegradation of some pesticides can vary widely as a function of site conditions. As stated by Howard (1991) in discussing 2,4-D in surface waters,

> the rate will depend on a number of factors including presence of acclimated organisms, nutrient levels, temperature and concentration of 2,4-D. Half-lives in river water of 18 to over 50 days (clear water) and 10 to 25 days (muddy water) with lag times of 6 to 12 days have been reported....Degradation is poor in oligotrophic water and where high 2,4-D concentrations are present and 2,4-D was not mineralized in water from 2 or 3 lakes tested.

Clearly, environmental conditions (such as temperature and oxic or anoxic conditions in soil or sediment), nutrient availability, and factors controlling pesticide speciation (dissolved vs sorbed) can greatly affect biodegradation. Perhaps the general uncertainty in biodegradation rates is best captured in the tendency of some investigators to refer simply to individual pesticides as "non-persistent," "moderately persistent," and "highly persistent." For example, Corbin et al. (2004) describe 2,4-D as "non-persistent" ($t_{1/2}$ = 6.2 days) in terrestrial environments, "moderately persistent" ($t_{1/2}$ = 45 days) in aerobic aquatic environments, and "highly persistent" ($t_{1/2}$ = 231 days) in anaerobic terrestrial and aquatic systems.

An Example of Current Modeling Input Choices in the Face of Parameter Uncertainty

To understand the approaches being used to account for uncertainty in modeling parameters, one can consider how biodegradation information was used in a PRZM-EXAM analysis of the ethyl hexyl ester (EHE) of 2,4-D (see Table 3-3). The compound is a nonionic ester, is quite hydrophobic, and thus is

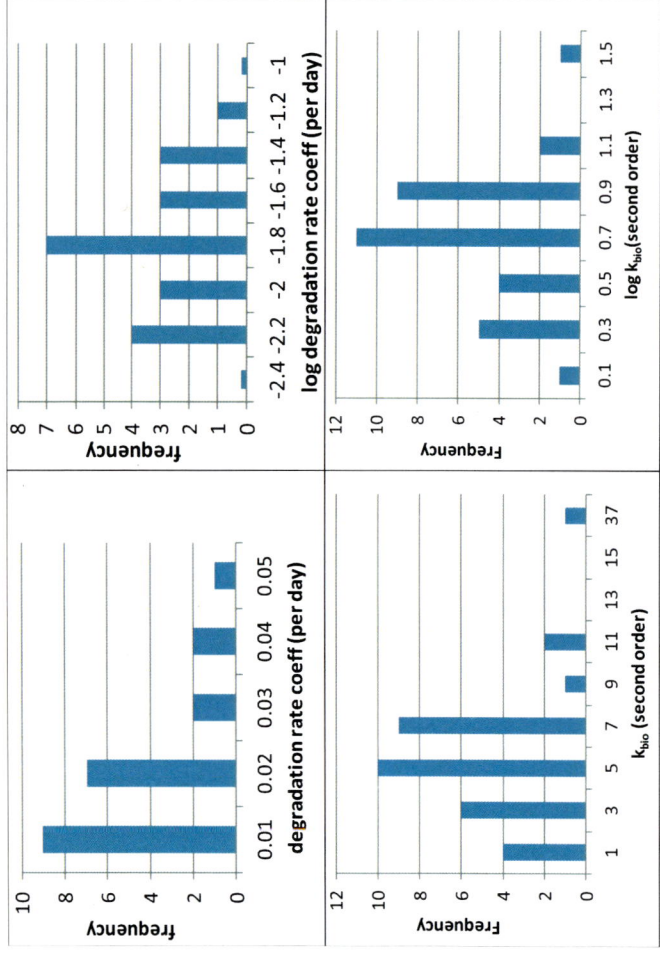

FIGURE 3-3 Upper panels: Distribution of observed (pseudo-) first-order biodegradation rates (per day) of flumetsulam as reported for 21 test soils by Lehman et al. (1992) on linear (left) and logarithmic (right) scales. Lower panels: Distribution of observed bacterial-number-normalized biodegradation rates (L/organism-hour) of the butoxyethyl ester of 2,4-D as reported for 33 test surface waters by Paris et al. (1981) on linear (left) and logarithmic (right) scales.

Exposure 77

highly sorptive (K_{oc} about 10,500). In a typical soil with an organic carbon content near 1%, one expects a sorption coefficient near 100 L/kg. That value means that almost all the ester (over 99%) is sorbed and somewhat unavailable to microorganisms. The extensive sorption implies that soil-to-soil differences in organic carbon content will affect the ester's bioavailability correspondingly and change the biodegradation rate accordingly. For example, soil with 3% organic carbon content will limit this ester's bioavailability by a factor of 3 relative to soil with 1% organic carbon content.

Next, the aerobic soil metabolism rate listed in Table 3-3 is based on a single soil-volatility study in which the ester was seen to degrade with a half-life of 8 days. Clearly, that information is not enough to provide any sense of the statistical variability in the biodegradation rate. Consequently, Corbin et al. (2004) compensated by providing some margin of safety, cutting the rate by a factor of 3 to arrive at a half-life of 24 days in aerobic soil. However, no scientific justification for a factor of 3 is provided; such a choice would require more observations. Furthermore, as directed in the modeling guidance (see Footnote b of Table 3-3), the aerobic aquatic degradation rate was set to half the value used for the aerobic soil case, yielding a half-life of 48 days. In the absence of any data, the guidance also leads one to assume that the ethylhexyl ester of 2,4-D is "stable" in anaerobic medias, such as sediments. Given those somewhat arbitrary inputs, PRZM/EXAM proceeds to estimate environmental concentrations of 2,4-D EHE. The approach leaves no possibility of assessing the probability distributions of the resultant EECs. The outcome of the model is quantitative, but its accuracy and precision are unknown.

Quantifying Parameter Uncertainty

It is clear from the above discussion that input parameters for fate and transport models have several components of uncertainty, including differences associated with environmental variability, imprecision of measurements under natural conditions, and lack of knowledge (see Chapter 2). Therefore, use of single values in a deterministic modeling approach provides an unwarranted sense of accuracy in predicting pesticide fate and later concentrations in water or loading on sediments and soils. As discussed in Chapter 2, the committee recommends taking a probabilistic approach and assigning appropriate distributions to the input parameters instead of single values. The committee notes that EPA has been working on probabilistic exposure modeling for many years (see, for example, Burns 2001). Model runs can be done with Monte Carlo techniques as single-level or multilevel models. The output is then a range of possible environmental concentrations with their associated probabilities of occurrence. That approach provides the required input information for comparison with hazard function to provide a probabilistic risk estimate.

TABLE 3-3 Biodegradation Rate Coefficients and Other Physical-Chemical Data Used in PRZM/EXAMS Fate Modeling of the Ethylhexyl Ester of 2,4-D

Model Parameter	Value	Comments	Source
Aerobic soil metabolism, $t_{1/2}$	24 days[a]	Estimated upper 90th percentile	MRID 42059601
Aerobic aquatic degradation, $t_{1/2}$ (KBACW)	48 days	Half the aerobic soil metabolism degradation rate	Estimated per EFED Guidance[b]
Anaerobic aquatic degradation, $t_{1/2}$ (KBACS)	Stable	No data	Estimated per EFED Guidance[2]
Aqueous photolysis, $t_{1/2}$	128 days		MRID 42749702
Hydrolysis, $t_{1/2}$	48 days		MRID 42735401
K_{oc}	10,500 mL/g		Estimated by EpiSuite Software
Molecular weight	333.26		Product Chemistry
Water solubility	0.32 mg/L		Product Chemistry
Vapor pressure	4.57 E-6 mm Hg		Product Chemistry
Henry's law constant	5.78 E-5 atm-m^3/mole		Product Chemistry

[a]Three times (upper 90th percentile) based on single soil half-life estimated from acceptable laboratory volatility study of 8 days.
[b]From *Guidance for Chemistry and Management Practice Input Parameters for Use in Modeling the Environmental Fate and Transport of Pesticides*, dated February 28, 2002.
Abbreviations: EFED, Environmental Fate and Effects Division of EPA; KBACS, first-order rate constant for pesticide's bacterial degradation in sediment (day^{-1}); KBACW, first-order rate constant for pesticide's bacterial degradation in water (day^{-1}); MRID, master record identification number (a unique cataloging number assigned to an individual pesticide study at the time of its submission to the agency).
Source: Corbin et al. 2004.

Interdependence of Input Parameters

Intervariable dependence can result in large uncertainty in model output, particularly when probabilistic modeling techniques are used. Assuming that all variables in an assessment are mutually independent will lead to erroneous risk results that might be conservative. That situation occurs whether the distributions characterizing the several variables represent natural variability or lack of information. Correctly modeling dependences usually requires additional empirical information beyond means, dispersions, and marginal distributions of input

Exposure

parameters and requires special modeling and mathematical techniques to propagate dependence. It seems unlikely that all input parameters are independent or perfectly dependent; these are the cases for which relatively simple solutions exist. However, the complexities of modeling incomplete dependence of multiple model parameters (Ferson et al. 2004; Kurowicka and Cooke 2006) probably outweigh the increased precision of the models. Therefore, for regulatory purposes, including ESA consultations, the committee recommends continuing to use the simplifying assumption of independence of model input parameters and to acknowledge the residual uncertainties of such an approach.

CONCLUSIONS AND RECOMMENDATIONS

Exposure-Modeling Practices

- Although the mass-balance models have many strengths, model limitations need to be recognized, and the appropriate model needs to be used for different risk-assessment contexts. Accordingly, a screening-level model should not be used when a refined exposure analysis is needed, such as in Step 2 or 3 assessments in the ESA process.
- To estimate pesticide exposure concentrations at various stages, the committee proposes a stepwise approach to exposure modeling. Step 1 would determine whether a pesticide and listed species overlap geographically and temporally. Step 2 would first identify the most important fate processes and other related considerations and then simplify the pesticide-fate model to estimate time-varying and space-varying pesticide concentrations in generic habitats relevant to the listed species. Step 3 would use refined models and the regional-specific or site-specific input values relevant to the listed species.
- Field studies need to be distinguished from general monitoring studies that are not associated with specific pesticide applications under well-described conditions. The latter cannot be used to estimate pesticide concentrations after a pesticide application or to evaluate model performance.
- Model predications are only as accurate or precise as parameter information. Thus, key processes need to be identified and the associated parameter values well defined.

Geospatial Data

- Although data on species occurrence inevitably are incomplete, uncertainties in modeled distributions of species typically can be quantified, and statistical characterizations of species distributions, species-environment relations, and the location and quality of habitat are more objective and reliable than qualitative descriptions of habitat.

- Existing and authoritative geospatial data on many scales are sufficient to support a substantial majority of habitat delineations and exposure analyses under the ESA and FIFRA. Widely recognized sources of data on environmental attributes—including topography, hydrography, meteorology, solar radiation, soils, geology, and land cover—can be used reliably for modeling species distributions and chemical fate. The authoritative sources that are most useful will vary among species and pesticides.

- Use of data and metadata that comply with the National Spatial Data Infrastructure can increase the clarity and repeatability of data analysis; facilitate quantification or even reduction of uncertainties in analytical results; and improve communication.

Uncertainties

- Any exposure analysis involving pesticide applications should at least qualitatively describe the potential effect of inerts on the environmental fate of an active ingredient. If the available information suggests that inerts (or adjuvants) might substantially affect the fate or transport of an active ingredient, the effect should be assessed quantitatively if data to support such a consideration are available.

- The extent to which the environmental fate of inerts or adjuvants needs to be considered quantitatively will depend largely on toxicological considerations (see Chapter 4). In the absence of information on the environmental-fate properties of inerts or adjuvants, quantitative structure-activity relationships can be used to estimate fate properties, but the use of such estimates will add to the uncertainties in the exposure analysis.

- Ideally, any risk assessment or BiOp should be based on exposures to pesticide components and other chemical agents that will occur in the field. Nonetheless, few methods are available for assessing exposure to environmental mixtures quantitatively or for predicting the relative concentrations of different mixture components in various environmental media, especially water and sediments. Monitoring data on the pesticides and other stressors will provide information about what is occurring in a specific area of concern but are not useful for model comparisons.

- In the absence of quantitative estimates of exposure, assessors should exclude potential mixture components from quantitative assessments. Uncertainties associated with the identities or exposure concentrations of potential mixture constituents should be qualitatively described to a decision-maker.

- Many diverse parameters are used in chemical-fate models, and their accuracy is important ultimately for the concentrations estimated in modeling efforts. However, little effort has been expended to evaluate the date inputs relevant to particular ESA evaluations. Therefore, if the agencies want to obtain more accurate modeling results, a subset of case-specific exposure estimates

Exposure

should be evaluated by pursing a measurement campaign specifically coordinated with several pesticide field applications.

- Sorption and biodegradation are important chemical-fate processes that are often associated with substantial uncertainty or represented inaccurately in fate models. More sorption data are needed to characterize nonlinear isotherms over concentration ranges and under conditions that are applicable to relevant agricultural settings, such as pH, ionic composition, and solid-phase mineralogy. Likewise, more data are needed to determine biodegradation coefficients, whether biodegradation rates are normally or log-normally distributed, and under which circumstances lag periods are important.

REFERENCES

Accardi-Dey, A.M., and P.M. Gschwend. 2002. Assessing the combined roles of natural organic matter and black carbon as sorbents in sediments. Environ. Sci. Technol. 36(1):21-29.

Araújo, M.B., and R.G. Pearson. 2005. Equilibrium of species' distributions with climate. Ecography 28(5):693-695.

ATSDR (Agency for Toxic Substances and Disease Registry). 1999. Toxicological Profile for Total Petroleum Hydrocarbons (TPH). U.S. Department of Health and Human Services, Public Health Service, Agency for Toxic Substances and Disease Registry, Atlanta, GA. September 1999 [online]. Available: http://www.atsdr.cdc.gov/ToxPro files/tp.asp?id=424&tid=75 [accessed Feb. 25, 2012].

Basham, G.W., and T.L. Lavy. 1987. Microbial and photolytic dissipation of imazaquin in soil. Weed Sci. 35(6):865-870.

Beestman, G.B. 1996. Emerging technology: The bases for new generations of pesticide formulations. Pp. 43-68 in Pesticide Formulation and Adjuvant Technology, C.L. Foy, and D.W. Pritchard, eds. Boca Raton, FL: CRC Press.

Belden, J.B., R.J. Gilliom, J.B. Martin, and M.J. Lydy. 2007. Relative toxicity and occurrence patterns of pesticide mixtures in streams draining agricultural watersheds dominated by corn and soybean production. Integr. Environ. Assess. Manag. 3(1): 90-100.

Bird, S.L., S.G. Perry, S.L. Ray, and M.E. Teske. 2002. Evaluation of the AgDISP aerial spray algorithms in the AgDRIFT model. Environ. Toxicol. Chem. 21(3):672-681.

Boyce, M.S., P.R. Vernier, S.E. Nielsen, and F.K.A. Schmiegelow. 2002. Evaluating resource selection functions. Ecol. Model. 157(2-3):281-300.

Bradbury, R.B., R.A. Hill, D.C. Mason, S.A. Hinsley, J.D. Wilson, H. Balzter, G.Q.A. Anderson, M.J. Whittingham, I.J. Davenport, and P.E. Bellamy. 2005. Modeling relationships between birds and vegetation structure using airborne LiDAR data: A review with case studies from agricultural and woodland environments. Ibis 147(3): 443-452.

Brausch, J.M., and P.N. Smith. 2007. Toxicity of three polyethoxylated tallowamine surfactant formulations to laboratory and field collected fairy shrimp, *Thamnocephalus platyurus*. Arch. Environ. Contam. Toxicol. 52(2):217-221.

Buckland, S.T., D.R. Anderson, K.P. Burnham, J.L. Laake, D.L. Borchers, and L. Thomas. 2001. Introduction to Distance Sampling: Estimating Abundance of Biological Populations. New York: Oxford University Press.

Burns, L.A. 2001. Probabilistic Aquatic Exposure Assessment for Pesticides, 1. Foundations. EPA/600/R-01/071. U.S. Environmental Protection Agency, Washington, DC. September 2001.

Burns, L.A. 2004. Exposure Analysis Modeling System (EXAMS): User Manual and System Documentation. EPA/600/R-00/081, Revision G. National Exposure Research Laboratory, Office of Research and Development, U.S. Environmental Protection Agency, Washington, DC. May 2004 [online]. Available: http://www.epa.gov/ceam publ/swater/exams/exams29804/EXAMREVG.PDF [accessed Nov. 16, 2012].

Burns, L.A. 2006. User Manual for EXPRESS, the "EXAMS-PRZM Exposure Simulation Shell". EPA/600/R-06/095. NTIS PB2007-100140. Office of Research and Development, U.S. Environmental Protection Agency, Washington, DC. September 2006 [online]. Available: http://www.epa.gov/ceampubl/swater/express/Expr ess.pdf [accessed Nov. 16, 2012].

CA Department of Conservation. 2007a. Farmland Mapping & Monitoring Program [online]. Available: http://www.conservation.ca.gov/dlrp/fmmp/Pages/Index.aspx [accessed Nov. 14, 2012].

CA Department of Conservation. 2007b. Farmland Mapping & Monitoring Program-Important Farmland Map Categories [online]. Available: http://www.conservation. ca.gov/dlrp/fmmp/mccu/Pages/map_categories.aspx [accessed Nov. 14, 2012].

Clevenger, A.P., J. Wierzchowski, B. Chruszcz, and K.E. Gunson. 2002. GIS-generated, expert-based models for identifying wildlife habitat linkages and planning mitigation passages. Conserv. Biol. 16(2):503-514.

Corbin, M., W. Evans, and J. Hetrick. 2004. Environmental Fate and Effects Division's Risk Assessment for the Reregistration Eligibility Document for 2,4-Dichlorophenoxyacetic Acid (2,4-D). U.S. Environmental Protection Agency [online]. Available: http:// www.epa.gov/espp/litstatus/effects/24d/attachment-b.pdf [accessed Sept. 4, 2012].

DCMI (Dublin Core Metadata Initiative Limited). 2012. Metadata Basic [online]. Available: http://dublincore.org//metadata-basics/ [accessed Nov. 14, 2012].

Denis, M.H., and S. Delrot. 1997. Effects of salts and surfactants on foliar uptake and long distance transport of glyphosate. Plant Physiol. Biochem. 35(4):291-301.

Dickson, B.G., and P. Beier. 2007. Quantifying the influence of topographic position on cougar (*Puma concolor*) movement in southern California, USA. J. Zool. 271(3): 270-277.

Dragovich, J.D., R.L. Logan, H.W. Schasse, T.J. Walsh, W.S. Lingley, Jr., D.K. Norman, W.J. Gerstel, T.J. Lapen, J.E. Schuster, and K.D. Meyers. 2002. Geologic Map of Washington-Northwest Quadrant. Geologic Map GM-50. Washington State Department of Natural Resources [online]. Available: http://www.dnr.wa.gov/Publi cations/ger_gm50_geol_map_nw_wa_250k.pdf [accessed Nov. 14, 2012].

Effland, W.R., N.C. Thurman, and I. Kennedy. 1999. Proposed Methods For Determining Watershed-Derived Percent Cropped Areas and Considerations for Applying Crop Area Adjustments To Surface Water Screening Models. Presentation To FIFRA Science Advisory Panel, May 27, 1999 [online]. Available: http://www.epa.gov/ scipoly/sap/meetings/1999/may/pca_sap.pdf [accessed Mar. 13, 2013].

Elith, J., C.H. Graham, R.P. Anderson, M. Dudik, S. Ferrier, A. Guisan, R.J. Hijmans, F. Huettmann, J.R. Leathwick, A. Lehmann, J. Li, L.G. Lohmann, B.A. Loiselle, G. Manion, C. Moritz, M. Nakamura, Y. Nakazawa, J.M. Overton, T.A. Peterson, S.J. Phillips, K. Richardson, R. Scachetti-Pereira, R.E. Schapire, J. Soberon, S. Williams, M.S. Wisz, and N.E. Zimmermann. 2006. Novel methods improve prediction of species distributions from occurrence data. Ecography 29(2):129-151.

EPA (U.S. Environmental Protection Agency). 1993. Wildlife Exposure Factors Handbook. EPA/600/R-93/197. Office of Research and Development, U.S. Environmental Protection Agency, Washington, DC [online]. Available: http://cfpub.epa.gov/ncea/cfm/recordisplay.cfm?deid=2799#Download [accessed Mar. 11, 2013].

EPA (U.S. Environmental Protection Agency). 2001. GENEEC (Generic Estimated Environmental Concentrations): Development and Use of GENEEC Version 2.0 for Pesticides, Aquatic Ecological Exposure Assessment, May 1, 2001. Environmental Fate and Effects Division, Office of Pesticide Programs, U.S. Environmental Protection Agency, Washington, DC [online]. Available: http://www.epa.gov/oppefed1/models/water/geneec2_description.htm [accessed Mar. 13, 2013].

EPA (U.S. Environmental Protection Agency). 2010. Label Review Manual [online]. Available: http://www.epa.gov/oppfead1/labeling/lrm/ [accessed Nov. 12, 2012].

EPA (U.S. Environmental Protection Agency). 2011. Inert Ingredients Permitted for Use in Nonfood Use Pesticide Products. Office of Prevention, Pesticides, and Toxic Substances, U.S. Environmental Protection Agency, Washington, DC. April 2011 [online]. Available: http://www.epa.gov/opprd001/inerts/inert_nonfooduse.pdf [accessed Nov. 12, 2012].

EPA (U.S. Environmental Protection Agency). 2012a. Technical Overview of Ecological Risk Assessment Risk Characterization. U.S. Environmental Protection Agency [online]. Available: http://www.epa.gov/oppefed1/ecorisk_ders/toera_risk.htm [accessed Mar. 22, 2013].

EPA (U.S. Environmental Protection Agency). 2012b. PRZM Version Index. Exposure Assessment Models Groundwater, U.S. Environmental Protection Agency [on-line]. Available: http://www.epa.gov/ceampubl/gwater/przm3/ [accessed Nov. 14, 2012].

EPA (U.S. Environmental Protection Agency). 2012c. EXAMS Version Index. Exposure Assessment Models Surface Water, U.S. Environmental Protection Agency [on-line]. Available: http://www.epa.gov/ceampubl/swater/exams/ [accessed Nov. 14, 2012].

EPA (U.S. Environmental Protection Agency). 2012d. Terrestrial Models. U.S. Environmental Protection Agency [online]. Available: http://www.epa.gov/oppefed1/models/terrestrial/ [accessed Mar. 22, 2013].

EPA (U.S. Environmental Protection Agency). 2012e. Surf Your Watershed. U.S. Environmental Protection Agency [online]. Available: http://cfpub.epa.gov/surf/locate/index.cfm [accessed Feb. 29, 2012].

EPA (U.S. Environmental Protection Agency). 2012f. GCSOLAR. Exposure Assessment Models Surface Water, U.S. Environmental Protection Agency [online]. Available: http://www.epa.gov/ceampubl/swater/gcsolar/ [accessed Nov. 14, 2012].

EPA (U.S. Environmental Protection Agency). 2012g. Pesticide Inert ingredients [online]. Available: http://www.epa.gov/opprd001/inerts/[accessed Nov. 14, 2012].

EPA (U.S. Environmental Protection Agency). 2012h. InertFinder [online]. Available: http://iaspub.epa.gov/apex/pesticides/f?p=901:1:1130545917893601 [accessed Nov. 16, 2012].

EPA (U.S. Environmental Protection Agency). 2012i. EPA Response to NAS Questions, March 28, 2012 [online]. Available: http://www.thecre.com/forum1/wp-content/uploads/2012/04/ESAEPAATRAZINE.pdf [accessed Nov. 15, 2012].

ESRI. 2010. Topography Tools for ArcGIS (9.3, 9.2, 9.1/9.0). ESRI, Redlands, CA [online]. Available: http://arcscripts.esri.com/details.asp?dbid=15996 [accessed Nov. 14, 2012].

ESRL (Earth System Research Laboratory). 2012. PSD Gridded Climate Datasets. National Oceanic and Atmospheric Administration, Earth System Research Laboratory [online]. Available: http://www.esrl.noaa.gov/psd/data/gridded/ [accessed Nov. 14, 2012].

Ferrell, J.A., G.E. MacDonald, and B. Sellers. 2008. Adjuvants. Document SS-AGR-109. Electronic Data Information Source (EDIS), University of Florida (UF), Institute of Food and Agricultural Sciences (IFAS) [online]. Available: http://edis.ifas.ufl.edu/wg050 [accessed Feb. 26, 2012].

Ferson, S., R.B. Nelsen, J. Hajagos, D.J. Berleant, J. Zhang, W.T. Tucker, L.R. Ginzburg, and W.L. Oberkamp. 2004. Dependence in Probabilistic Modeling, Dempster-Shafer Theory, and Probability Bounds Analysis. SAND2004-3072. Sandia National Laboratories, Albuquerque, NM [online]. Available: http://www.ramas.com/depend.pdf [accessed Mar. 25, 2013].

FGDC (Federal Geographic Data Committee). 2007. National Spatial Data Infrastructure [online]. Available: http://www.fgdc.gov/nsdi/nsdi.html [accessed Mar. 7, 2012].

FGDC (Federal Geographic Data Committee). 2012. Geospatial Metadata. What are Metadata? [online]. Available: http://fgdc.gov/metadata [accessed Nov. 14, 2012].

Fletcher, J.S., J.E. Nellesson, and T.G. Pfleeger. 1994. Literature review and evaluation of the EPA food-chain (Kenaga) nomogram, an instrument for estimating pesticide residues on plants. Environ. Toxicol. Chem. 13(9):1383-1391.

Franklin, J. 2009. Mapping Species Distributions: Spatial Inference and Prediction. Cambridge, UK: Cambridge University Press.

Fretwell, S.D. 1972. Populations in a Seasonal Environment. Princeton, NJ: Princeton University Press.

Gerstl, Z. 1990. Estimation of organic chemical sorption by soils. J. Contam. Hydrol. 6(4): 357-375.

Getzin, L.W. 1970. Persistence of methidathion in soils. Bull. Environ. Contam Toxicol. 5(2):104-110.

Getzin, L.W. 1973. Persistence and degradation of carbofuran in soils. Environ. Entomol. 2(3):461-468.

Getzin, L.W., and I. Rosefield. 1966. Persistence of diazinon and zinophos in soils. J. Econ. Entomol. 59(3):512-516.

Gilliom, R.J., J.E. Barbash, C.G. Crawford, P.A. Hamilton, J.D. Martin, N. Nakagaki, L.H. Nowell, J.C. Scott, P.E. Stackelberg, G.P. Thelin, and D.M. Wolock. 2007. The Quality of Our Nation's Waters: Pesticides in the Nation's Streams and Groundwater, 1992-2001. U.S. Geological Survey Circular 1291. U.S. Geological Survey, Reston, VA [online]. Available: http://pubs.usgs.gov/circ/2005/1291/pdf/circ1291.pdf [accessed Nov. 13, 2012].

Hall, L.S., P.R. Krausman, and M.L. Morrison. 1997. The habitat concept and a plea for standard terminology. Wildl. Soc. Bull. 25(1):173-182.

Han, W., Z. Yang, L. Di, and R. Mueller. 2012. CropScape: A Web service based application for exploring and disseminating U.S. conterminous geospatial cropland data products for decision support. Comput. Electron. Agr. 84:111-123.

Harvey, J., Jr., and H.L. Pease. 1973. Decomposition of the methomyl in soil. J. Agric. Food Chem. 21(5):784-786.

Hoerger, F., and E.E. Kenaga. 1972. Pesticide residues on plants: Correlation of representative data as a basis for estimation of their magnitude in the environment. Pp. 9-28 in Environmental Quality and Safety: Chemistry, Toxicology, and Technology, Vol. 1. Global Aspects of Chemistry, Toxicology, and Technology as Applied to the Environment, F. Coulston, and F. Korte, eds. Stuttgart: Georg Thieme Verlag.

Hoogeweg, C.G., W.M. Williams, R. Breuer, D. Denton, B. Rook, and C. Watry. 2011. Spatial and Temporal Quantification of Pesticide Loadings to the Sacramento River, San Joaquin River, and Bay-Delta to Guide Risk Assessment for Sensitive Species. CALFED Science Grant No. 1055. Waterborne Environmental, Inc., Leesburg, VA, and University of California, Davis, CA. November 2, 2011. 293 pp [online]. Available: http://www.waterborne-env.com/publications/papers_reports/CALFED_Final_Report_2011-Nov_2.pdf [accessed on May 9, 2012].

Hope, B.K. 2012. Using legacy data to relate biological condition to cumulative aquatic toxicity in the Willamette River Basin (Oregon, USA). Arch. Environ. Contam. Toxicol. 62(3):424-437.

Horning, N., J.A. Robinson, E.J. Sterling, W. Turner, and S. Spector. 2010. Remote Sensing for Ecology and Conservation: Handbook of Techniques. New York: Oxford University Press.

Howard, P.H., ed. 1991. 2,4-D. Pp. 145-156 in Handbook of Environmental Fate and Exposure Data for Organic Chemicals. Boca Raton, FL: CRC Press.

Hyun, S., and L.S. Lee. 2005. Quantifying the contribution of different sorption mechanisms for 2,4-dichlorophenoxyacetic acid sorption by several variable-charge soils. Environ. Sci. Technol. 39(8):2522-2528.

Jenness Enterprises. 2011. DEM Surface Tools. Jenness Enterprises, Flagstaff, AZ [online]. Available: http://www.jennessent.com/arcgis/surface_area.htm [accessed Nov. 14, 2012].

Karl, H.A., L.E. Susskind, and K.H. Wallace. 2007. A dialogue, not a diatribe: Effective integration of science and policy through joint fact finding. Environment 49(1):20-34.

Katagi, T. 2008. Surfactant effects on environmental behavior of pesticides. Rev. Environ. Contam. Toxicol. 194:71-177.

Knisel, W.G., and F.M. Davis. 2000. GLEAMS (Groundwater Loading Effects of Agricultural Management Systems), Version 3.0, User Manual. Publication No. SEWRL-WGK/FMD-050199. U.S. Department of Agriculture, Agricultural Research Service, Southeast Watershed Research Laboratory, Tifton, GA. August 15, 2000. 194pp.

Konrad, J.G., and G. Chesters. 1969. Degradation in soil of ciodrin, an organophosphate insecticide. J. Agric. Food Chem. 17(2):226-230.

Kosswig, K. 1994. Surfactants. Pp. 747-815 in Ullmann's Encyclopedia of Industrial Chemistry, Vol. 25A, Starch and Other Polysaccharides to Surfactants, 5th Ed. Weinheim: Wiley-VCH.

Kurowicka, D., and R.M. Cooke. 2006. Uncertainty Analysis with High Dimensional Dependence Modeling. Hoboken, NJ: John Wiley & Sons Inc.

Laskowski, D.A. 1995. EPA guidelines for environmental fate studies: Meaningful data for assessing exposure to pesticides. Pp. 117-128 in Agrochemical Environmental Fate: State of the Art, M.L. Leng, E.M.K. Leovey, and P.P. Zubkoff, eds. New York: Lewis.

Laskowski, D.A., and A.J. Regoli. 1972. Influence of the environment on N-SERVE stability. Agron. Abstacts 97 (as cited in Laskowski 1995).

Laskowski, D.A., C.A.I. Goring, P.J. McCall, and R.L. Swann. 1982. Terrestrial environment. Pp.198-240 in Environmental Risk Analysis for Chemicals, R.C. Conway, ed. New York: Van Nostrand Reinhold.

Lehmann, R.G., J.R. Miller, D.D. Fontaine, D.A. Laskowski, J.H. Hunter, and R.C. Cordes. 1992. Degradation of a sulfonamide herbicide as a function of soil sorption. Weed Res. 32(3):197-205.

Leonard, R.A., W.G. Knisel, and D.A. Still. 1989. GLEAMS: Groundwater loading effects of agricultural management systems. Transactions ASAE. 30:1403-1418.

Lin, K., D. Haver, L. Oki, and J. Gan. 2009. Persistence and sorption of fipronil degradates in urban stream sediments. Environ. Toxicol. Chem. 28(7):1462-1468.

Liu, Z. 2004. Effects of surfactants on foliar uptake of herbicides - a complex scenario. Colloids Surf. B Biointerfaces 35(3-4):149-153.

Loague, K., and R.E. Green. 1991. Statistical and graphical methods for evaluating solute transport models: Overview and application. J. Contam. Hydrol. 7:51-73.

Lode, O. 1967. Decomposition of linuron in different soils. Weed Res. 7(3):185-190.

MacKenzie, D.I., J.D. Nichols, J.A. Royle, K.H. Pollock, L.L. Bailey, and J.E. Hines. 2006. Occupancy Estimation and Modeling: Inferring Patterns and Dynamics of Species Occurrence. San Diego, CA: Academic Press.

Manly, B.F., L. McDonald, D.L. Thomas, T.L. MacDonald, and W.P. Erickson. 2010. Resource Selection by Animals: Statistical Design and Analysis for Field Studies, 2nd Ed. Dordrecht, the Netherlands: Kluwer.

Martinuzzi, S., L.A. Vierling, W.A. Gould, M.J. Falkowski, J.S. Evans, A.T. Hudak, and K.T. Vierling. 2009. Mapping snags and understory shrubs for a LiDAR-based assessment of wildlife habitat suitability. Remote Sens. Environ. 113(12):2533-2546.

Mitchell, S.C. 2005. How useful is the concept of habitat? A critique. Oikos 110(3):634-638.

Morrison, M.L., and L.S. Hall. 2002. Standard terminology: Toward a common language to advance ecological understanding and application. Pp. 43-52 in Predicting Species Occurrences: Issues of Accuracy and Scale, J.M. Scott, P.J. Heglund, M.L. Morrison, J.B. Haufler, M.G. Raphael, W.A. Wall, and F.B. Samson, eds. Covelo, CA: Island Press.

NASS (National Agricultural Statistics Service). 2010. Charts and Maps. U.S. Department of Agriculture, National Agricultural Statistics Service [online]. Available: http://www.nass.usda.gov/Charts_and_Maps/index.asp [accessed Nov. 14, 2012].

Nagy, K.A., and C.C. Peterson. 1988. Scaling of Water Flux Rates in Animals. Berkeley, CA: University of California Press.

NCDC (National Climatic Data Center). 2012. NOAA Regional Climate Centers [online]. Available: http://www.ncdc.noaa.gov/oa/climate/regionalclimatecenters.html [accessed Nov. 14, 2012].

NMFS (National Marine Fisheries Service). 2008. Biological Opinion, Environmental Protection Agency Registration of Pesticides Containing Chlorpyrifos, Diazinon, and Malathion. National Marine Fisheries Service, Silver Spring, MD. November 18, 2008 [online]. Available: http://www.nmfs.noaa.gov/pr/pdfs/pesticide_biop.pdf [accessed Nov. 13, 2012].

NMFS (National Marine Fisheries Service). 2009. Biological Opinion, Endangered Species Act Section 7 Consultation, Environmental Protection Agency Registration of Pesticides Containing Carbaryl, Carbofuran, and Methomyl. National Marine Fisheries Service, Silver Spring, MD. April 20, 2009 [online]. Available: http://www.nmfs.noaa.gov/pr/pdfs/carbamate.pdf [accessed Nov. 12, 2012].

NMFS (National Marine Fisheries Service). 2010. Biological Opinion, Endangered Species Act Section 7 Consultation, Environmental Protection Agency Registration of Pesticides Containing Azinphos methyl, Bensulide, Dimethoate, Disulfoton, Ethoprop, Fenamiphos, Naled, Methamidophos, Methidathion, Methyl parathion, Phorate and Phosmet. National Marine Fisheries Service, Silver Spring, MD. August 31, 2010 [online]. Available: http://www.nmfs.noaa.gov/pr/pdfs/final_batch_3_opinion.pdf [accessed Nov. 13, 2012].

NMFS (National Marine Fisheries Service). 2011. Biological Opinion, Endangered Species Act Section 7 Consultation, Environmental Protection Agency Registration of Pesticides 2,4-D, Triclopyr BEE, Diuron, Linuron, Captan, and Chlorothalonil. National Marine Fisheries Service, Silver Spring, MD. June 30, 2011 [online]. Available: http://www.nmfs.noaa.gov/pr/pdfs/consultations/pesticide_opinion4.pdf [accessed Nov. 13, 2012].

Noon, B.R. 1981. Techniques for sampling avian habitats. Pp. 42-52 in The Use of Multivariate Statistics in Studies of Wildlife Habitat, D.E. Capen, ed. General Technical Report RM-89. U. S. Forest Service, Rocky Mountain Forest and Range Experimental Station, Fort Collins, CO [online]. Available: http://fishwild.vt.edu/Chihuahua%202005/READINGS/noon.pdf [accessed Nov. 13, 2012].

Novak, J.M., T.B. Moorman, and C.A. Cambardella. 1997. Atrazine sorption at the field scale in relation to soils and landscape position. J. Environ. Qual. 26(5):1271-1277.

NRC (National Research Council). 1993. Toward a Coordinated Spatial Data Infrastructure for the Nation. Washington, DC: National Academy Press.

NRCS (National Resource Conservation Service). 2012a. Watershed Boundary Data Set [online]. Available: http://www.nrcs.usda.gov/wps/portal/nrcs/main/national/water/watersheds/dataset [accessed Feb. 29, 2012].

NRCS (National Resource Conservation Service). 2012b. Web Soil Survey (WSS) [online]. Available: http://websoilsurvey.nrcs.usda.gov/app/HomePage.htm [accessed Nov. 14, 2012].

NREL (National Renewable Energy Laboratory). 2012. SMARTS: Simple Model of the Atmospheric Radiative Transfer of Sunshine [online]. Available: http://www.nrel.gov/rredc/smarts/ [accessed Nov. 14, 2012].

Osborne, P.E., J.C. Alonso, and R.G. Bryant. 2001. Modeling landscape-scale habitat use using GIS and remote sensing: A case study with great bustards. J. Appl. Ecol. 38(2):458-471.

Overwatch Systems, LTD. 2009. VLS Software: LIDAR Analyst® [online]. Available: http://www.vls-inc.com/lidar_analyst.htm [accessed Nov. 14, 2012].

Paris, D.F., W.C. Steen, G.L. Baughman, and J.T. Barnett, Jr. 1981. Second-order model to predict microbial degradation of organic compounds in natural waters. Appl. Environ. Microbiol. 41(3): 603-609.

Pearson, R.G., and T.P. Dawson. 2003. Predicting the impacts of climate change on the distribution of species: Are bioclimate envelope models useful? Global Ecol. Biogeogr. 12(5):361-371.

Pelletier, J. 2008. Quantitative Modeling of Earth Surface Processes. Cambridge, UK: Cambridge University Press.

Rich, P.M., W.A. Hetrick, and S.C. Saving. 1994. Modeling Topographic Influences on Solar Radiation: A Manual for the SOLARFLUX Model. LA-12989-M. Los Alamos National Laboratory, Los Alamos, NM [online]. Available: http://ntrs.nasa.gov/archive/nasa/casi.ntrs.nasa.gov/19960048095_1996066639.pdf [accessed Nov. 13, 2012].

Richey, F.A., Jr., W.J. Bartley, and K.P. Sheets. 1977. Laboratory study on the degradation of (the pesticide) aldicarb in the soils. J. Agric. Food. Chem. 25(1):47-51.

Ritter, D.F., R.C. Kochel, and J.R. Miller. 2011. Process Geomorphology. Long Grove, IL: Waveland Press Inc.

Royle, J.A., R.B. Chandler, C. Yackulic, and J.D. Nichols. 2012. Likelihood analysis of species occurrence probability from presence-only data for modeling species distributions. Method Ecol. Evol. 3(3):545-554.

Ruepell, M.L., B.B. Brightwell, J. Schaefer, and J.T. Marvel. 1977. Metabolism and degradation of glyphosate in soil and water. J. Agric. Food Chem. 25(3):517-528.

Samways, M.J., M.A. McGeoch, and T.R. New. 2010. Insect Conservation: A Handbook of Approaches and Methods. Oxford, UK: Oxford University Press.

Sato, C., and J.L. Schnoor. 1991. Applications of three completely mixed compartment models to the long-term fate of dieldrin in a reservoir. Water Res. 25(6):621-631.

Schwartz, M.W. 2012. Using niche models with climate projections to inform conservation management decisions. Biol. Conserv. 155:149-156.

Scott, J.M., F. Davis, B. Csuti, R. Noss, B. Butterfield, C. Groves, H. Anderson, S. Caicco, F. D'Erchia, T.C. Edwards, Jr., J. Ulliman, and R.G. Wright. 1993. Gap Analysis: A Geographical Approach to Protection of Biological Diversity. Wildlife Monographs 123. Lawrence, KS: Allen Press.

Scott, J.M., P.J. Heglund, M.L. Morrison, J.B. Haufler, M.G. Raphael, W.A. Wall, and F.B. Samson, eds. 2002. Predicting Species Occurrences: Issues of Accuracy and Scale. Covelo, CA: Island Press.

SDTF (Spray Drift Task Force). 2010. Important Information about AgDRIFT® Model [online]. Available: http://www.agdrift.com/AgDRIFt2/DownloadAgDrift2_0.htm [accessed Nov. 14, 2012].

Seaber, P.R., F.P. Kapinos, and G.L. Knapp. 1987. Hydrologic Unit Maps. U.S. Geological Survey Water-Supply Paper 2294. Washington, DC: U.S. Government Printing Office [online]. Available: http://pubs.usgs.gov/wsp/wsp2294/pdf/wsp_2294_a.pdf [accessed Nov. 13, 2012].

Seiber, J.N., M.M. McChesney, P.F. Sanders, and J.E. Woodrow. 1986. Models for assessing the volatilization of herbicides applied to flooded rice fields. Chemosphere 15(2):127-138.

Seth, R., D. Mackay, and J. Muncke. 1999. Estimating the organic carbon partition coefficient and its variability for hydrophobic chemicals. Environ. Sci. Technol. 33(14):2390-2394.

Shriner, S.A, T.R. Simons, and G.L. Farnsworth. 2002. A GIS-based habitat model for Wood Thrush in Great Smoky Mountain National Park. Pp. 529-536 in Predicting Species Occurrences: Issues of Accuracy and Scale, J.M. Scott, P.J. Heglund, M.L. Morrison, J.B. Haufler, M.G. Raphael, W.A. Wall, and F.B. Samson, eds. Washington, DC: Island Press.

Sinclair, S.J., M.D. White, and G.R. Newell. 2010. How useful are species distribution models for managing biodiversity under future climates? Ecol. Soc. 15(1):8.

Skelly, D.K., L.N. Joseph, H.P. Possingham, L.K. Freidenburg, T.J. Farrugia, M.T. Kinnison, and A.P. Hendry. 2007. Evolutionary responses to climate change. Conserv. Biol. 21(5):1353-1356.

Stackelberg, P.E., L.J. Kauffman, M.A. Ayers, and A.L. Baehr. 2009. Frequently co-occurring pesticides and volatile organic compounds in public supply and monitoring wells, southern New Jersey, USA. Environ. Toxicol. Chem. 20(4):853-865.

Suarez, L.A. 2005. PRZM-3, A Model For Predicting Pesticide And Nitrogen Fate In The Crop Root And Unsaturated Soil Zones: User's Manual For Release 3.12.2. EPA/600/R-05/111. National Exposure Research Laboratory, Office of Research and Development, U.S. Environmental Protection Agency, Athens, GA [online]. Available: http://www.epa.gov/scipoly/sap/meetings/2008/october/przm.pdf [accessed Nov. 13, 2012].

Exposure 89

Teske, M.E., S.L. Bird, D.M. Esterly, T.B. Curbishley, S.L. Ray, and S.G. Perry. 2002. AgDRIFT: A model for estimating near-field spray drift from aerial applications. Environ. Toxicol. Chem. 21(3):659-671.

Tiedje, J.M., and B.B. Mason. 1974. Biodegradation of nitrilotriacetate (NTA) in soils. Soil Sci. Soc. Am. Proc. 38(2):278-283.

Tornero-Velez, R., E.P. Egeghy, and E.A. Cohen Hubal. 2012. Biogeographical analysis of chemical co-occurrence data to identify priorities for mixtures research. Risk Anal. 32(2):224-236.

Tu, M., and J.M. Randall. 2005. Adjuvants. Chapter 8 in Weed Control Methods Handbook: Tools and Techniques for Use in Natural Areas, M. Tu, C. Hurd, and J.M. Randall, eds. Nature Conservancy. Available: http://www.invasive.org/gist/products/handbook/21.Adjuvants.pdf [accessed Nov. 14, 2012].

U.S. Census Bureau. 2011. Statistical Abstract of the United States. Washington, DC: U.S. Government Printing Office [online]. Available: http://www.census.gov/prod/www/abs/statab2011_2015.html [accessed Feb. 29, 2012].

USGS (U.S. Geological Survey). 2010a. 1:250,000-scale Hydrologic Units of the United States. Water Resources NSDI [National Spatial Data Infrastructure] Node [online]. Available: http://water.usgs.gov/GIS/metadata/usgswrd/XMLhuc250k.xml [accessed Feb. 29, 2012].

USGS (U.S. Geological Survey). 2010b. Land Cover Data. The USGS Land Cover Institute (LCI) [online]. Available: http://landcover.usgs.gov/landcoverdata.php [accessed Nov. 14, 2012].

USGS (U.S. Geological Survey). 2011. National Gap Analysis Program (GAP) Species Data [online]. Available: http://gapanalysis.usgs.gov/species/ [accessed Feb. 19, 2013].

USGS (U.S. Geological Survey). 2012a. USGS Center for Lidar information Coordination and Knowledge [online]. Available: http://lidar.cr.usgs.gov/ [accessed Nov. 15, 2012].

USGS (U.S. Geological Survey). 2012b. Hydrologic Unit Maps. U.S. Geological Survey [online]. Available: http://water.usgs.gov/GIS/huc.html [accessed Feb. 29, 2012].

USGS (U.S. Geological Survey). 2012c. Geologic Maps of U.S. States. U.S. Geological Survey Mineral Resources [online]. Available: http://mrdata.usgs.gov/geology/state/ [accessed Nov. 14, 2012].

USGS/USDA/NRCS (U.S. Geological Survey, U.S. Department of Agriculture, and National Resource Conservation Service). 2011. Federal Standards and Procedures for the National Watershed Boundary Dataset (WBD), Chapter 3 of Section A, Federal Standards, Book 11. Collection and Delineation of Spatial Data. U.S. Geological Survey Techniques and Methods 11-A3, 2nd Ed. Reston, VA: U.S. Geological Survey [online]. Available: http://pubs.usgs.gov/tm/tm11a3/pdf/tm11-A3-Ed2.pdf [accessed Nov. 14, 2012].

Vierling, K.T., L.A. Vierling, W.A. Gould, S. Martinuzzi, and R.M. Clawges. 2008. LiDAR: Shedding new light on habitat characterization and modeling. Front. Ecol. Environ. 6(2):90-98.

WA Department of Ecology. 2012. Washington Water Resource Inventory Area (WRIA) Maps [online]. Available: http://www.ecy.wa.gov/services/gis/maps/wria/wria.htm [accessed Nov. 14, 2012].

Walker, A. 1976. Simulation of herbicide persistence in soil. III. Propyzamide in different soil types. Pestic. Sci. 7(1):59-64.

(FWS) and the National Marine Fisheries Service (NMFS)—collectively referred to as the Services—to make jeopardy determinations (Step 3, Figure 2-1).

SUBLETHAL, INDIRECT, AND CUMULATIVE EFFECTS

Pesticides can kill organisms that are closely or distantly related to their intended targets, and they can cause sublethal changes that can affect reproduction, shorten lifespans, or make the organisms unable to compete. The following sections discuss how to incorporate sublethal effects into ecological risk assessments, how effects on one organism might indirectly affect others, and how pesticide effects might be modified by exposure to other environmental stressors.

Sublethal Effects

Pesticides can have sublethal effects at multiple levels of biological organization: molecular, cellular, tissue, organism, population, and community. Only when compensatory or adaptive mechanisms at one level of biological organization begin to fail do deleterious effects become apparent at higher levels. The committee considered how to assess objectively the degree to which observed effects of pesticides on organisms are *adverse*. Defining that concept is essential for ecological risk assessment because even if an effect is reliably observed, that alone might not be sufficient to conclude that the effect is adverse. The committee concluded that the only reasonable way to determine whether an effect is adverse and how adverse it might be is to assess the degree to which it affects the organism's survival and reproductive success. It then is possible to extrapolate from changes in an individual organism's survival or reproductive success to estimate population effects. If an adverse effect is large enough, it might lead to extinction of the species. EPA reached a similar conclusion in its overview of the ecological risk-assessment process (EPA 2004, p. 31): "If the effects on the survival and reproduction of individuals are limited, it is assumed that the risk at the population level from such effects will be of minor consequence. However, as the risk of reductions in survival and/or reproduction rates increase, the greater the potential risk to populations."

EPA and the Services agree on the inclusion of sublethal effects in the risk-assessment process but disagree on the extent to which such effects should be included. For example, in its responses to committee questions, EPA explained that its focus is "on how to relate the relevance of sublethal data to an assessment of the risks to fitness of listed species," with *fitness* defined as "an individual's ability to survive and reproduce" (EPA 2012a, p. 2). Furthermore, EPA considers that incorporation of sublethal effects into an ecological risk assessment must be accompanied by an explicit relationship that defines the contribution of the sublethal effect to an individual organism's fitness in terms of the end points of "survival, growth and reproduction" (EPA 2012a, p. 20). EPA

Exposure 93

stated that it "does not believe that all sub-lethal effects or that all levels of a sub-lethal effect on an individual constitute a compromise of individual fitness" (EPA 2012a, p. 3).

EPA's approach differs from the Services' approach. For example, FWS "casts a wide net for each potentially affected species to ensure that the most sensitive endpoints are captured and evaluated" (FWS 2012, p. 2). It contends that "at present, data describing 'sub-lethal' effects are acknowledged but then set aside and not used by EPA in making effects determinations or characterizing the potential effects of the action, unless other data or studies are available that would enable EPA to quantify a relationship between the 'sub-lethal' effect and EPA's traditional endpoints, survival, growth, or reproduction." FWS (2012, pp. 2-3) continued that "in contrast, when characterizing the 'Effects of the Action' pursuant to the ESA [Endangered Species Act], the FWS does not limit itself to using only those data that quantify changes in survival, growth, or reproduction."

As discussed in the section on effects models below, assessing the effects of pesticides on listed species requires quantifying the effect of a pesticide on survival and reproduction of a species in the wild. Any effect that results in a change in one component is relevant to the assessment. In contrast, any effect that does not change either component is irrelevant with respect to a quantitative assessment of population effects. The relevance of any particular sublethal effect is likely to depend on the species. Growth, for example, might be a relevant effect in some species but not in others. In mammalian species, retarded growth might increase age of first reproduction but not affect reproductive output thereafter. In many fish species, size of the individual organism is directly related to reproductive output throughout the lifespan. Many plant species do not need to achieve a particular size for maximal reproductive output. Therefore, the committee recommends that EPA in Step 2 (see Figure 2-1) cast a wide net and identify information about sublethal effects of a chemical. If possible, EPA's assessment should include information about responses at various chemical concentrations (a concentration-response curve) and, at a minimum, include a qualitative assessment of the relationship between sublethal effects and survival and reproduction. In Step 3 (see Figure 2-1), the Services should show how such effects change demographic measures (survival or reproduction) of a listed species and incorporate such information into the population viability analyses or should state that such relationships are unknown but possible and include a qualitative discussion in the uncertainty section of the biological opinion (BiOp). The Services face the greatest challenge in Step 3 in determining whether an observed sublethal effect will change survival or reproduction in the natural population and, if so, the magnitude of such a change in relation to the predicted exposure.

Relationships between sublethal effects and changes in population growth rates span a continuum of uncertainty that depends on the ability to quantify the link. At one extreme, the relationship between a sublethal effect and survival or reproduction has not been quantified empirically, and the available mechanistic information is not sufficient to model the causal chain quantitatively. For example, markers of oxidative stress—such as glutathione or superoxide dismutase—indicate a physiological response to a chemical, but the relationship of the response to survival or reproduction is not known. Such a response could not be easily quantified with respect to population assessment if the observed response were the only pertinent information.

At the other extreme, the link between sublethal effects and population persistence might be clear, quantifiable, and well documented in the literature. For example, the singing ability of some male birds directly affects the probability of their establishing and holding a territory and forming pair bonds with mates (Spencer et al. 2003). Impaired singing ability could directly affect reproductive success during the breeding season if the male song did not attract a female mate. Similarly, impaired growth of juvenile salmon might result in a reduction in size of individual salmon as they migrate to sea and could reduce survival. Specifically, Baldwin et al. (2009) modeled the relationship between sublethal effects on acetylcholinesterase activity and feeding behavior of juvenile Chinook salmon and reductions in growth after short-term exposure to environmentally realistic concentrations of organophosphate and carbamate pesticides. Reductions in growth correlated with reduced size at ocean entry and with later survival. Mebane and Arthaud (2010) modeled the effects of sublethal effects of low concentrations of copper on growth of juvenile Chinook salmon and projected potential effects on population size, recovery rates, and extinction risks.

Many sublethal effects might have a link to population viability, but that link has not yet been quantified. An example is altered olfactory ability, which has been shown to increase predation risk in some species of salmon because of an inability to detect chemical cues that signal the presence of a predator or because of a loss of homing ability (Scholz et al. 2000). Whether altered olfactory ability affects survival will depend on the degree of its expression in the natural environment, the presence of predators during the time that olfaction is lost, and whether it occurs in fish whose size makes them susceptible to predation. Impaired immune function is another example in which an organism is affected, but the effect on population viability is unclear. A working immune system is critical for survival, but an alteration of some aspect of immune function and its effect on disease resistance are often less clear—for example, Does a given reduction in circulating leukocytes affect susceptibility to disease? Furthermore, the effect of an impaired immune system on disease susceptibility hinges partly on the presence of a pathogen. The committee notes that exposure to pesticides in some species might actually increase defense responses to predation. For example, Barry (1998) observed increased helmet formation—a defense response that deters predation efficiency—in daphnia exposed to low concentrations of endosulfan.

Exposure

Uncertainties in concentration-response relationships or differences between laboratory and field responses, particularly behavioral responses, further complicate the quantification of changes in survival and reproductive success in response to sublethal toxicity. Assessment of sublethal effects, as well as cumulative and indirect effects, is even more complicated in species that have complex life cycles and population structures, such as Pacific salmon (see Box 4-1 for further discussion).

The committee concludes that survival and reproduction are the principal effects in determining population viability. The inability to quantify the relationship between sublethal effects and survival or reproductive success does not negate the potential importance of such effects for population persistence. However, the relationship remains a hypothesis that can be described only qualitatively with reference to the scientific literature for why such a hypothesis is tenable. Implications for risk characterization can be discussed qualitatively, not quantitatively, as an additional uncertainty beyond uncertainties that are propagated in a formal quantitative manner. The narrative can be considered by a decision-maker according to the applicable policy constraints regarding risk tolerance. However, such a separation of important risk components and uncertainty into quantitative and qualitative portions that cannot formally be combined makes it difficult to integrate and interpret the results of a risk assessment. Integration can be improved by quantifying better the relationships that are viewed as critical for understanding the risks posed by a pesticide to a listed species. One way to facilitate integration of the hypothetical relationship into the formal risk assessment is to conduct extensive reviews of comparative data or empirical case studies or to conduct targeted new studies that could help to derive defensible scientific quantification of the links between sublethal effects and survival or reproduction.

Indirect Effects

The Services have defined *indirect effects* as "those that are caused by the proposed action and are later in time, but still are reasonably certain to occur" (50 C.F.R. 402.02). Thus, their definition from a regulatory standpoint characterizes indirect effects as simply delayed effects. Depending on how one interprets that definition, it could be quite restrictive and different from most ecologists' understanding of indirect effects, which typically include effects on prey, competitors, or predators of a listed species or on other aspects of the species' ecological milieu but not direct effects on the species. On the basis of the documents reviewed by the committee, it appears that the restrictive definition is not used by the agencies; therefore, this section discusses indirect effects as including those normally understood by the term.

BOX 4-1 Ecological Risk Assessment in Species That Have Complex Population Structure and Life History: Pacific Salmon and Trout

Pacific salmon and trout (*Oncorhynchus* spp.) are the basis of valuable commercial and recreational fisheries; part of the economy, ceremony, and subsistence of American Indians; components of complex ecosystems to which they contribute great quantities of nutrients; symbols of clean water and healthy rivers; and a host of other attributes related to human and natural systems (NRC 1996). Many factors have contributed to declines in salmon and trout, which are, in some cases, protected under the US Endangered Species Act (Gustafson et al. 2007). Protection of the listed distinct population segments (DPSs) has ramifications for a wide variety of human activities, including application of chemicals to control animals and plants that are considered crop pests and weeds.

There are five species of Pacific salmon in North America: Chinook, *O. tshawytscha*; coho, *O. kisutch*; sockeye, *O. nerka*; chum, *O. keta*; and pink, *O. gorbuscha*. There are also two trout species of the same genus: rainbow/steelhead trout, *O. mykiss*, and cutthroat trout, *O. clarkii*. Both trout species are quite variable phenotypically and have several subspecies (Behnke 1992). All Pacific salmon are spawned in freshwater, and most migrate to sea and return to freshwater at maturity to spawn (that is, they are anadromous); however, resident populations of sockeye salmon (kokanee) are well known and a few individuals of other salmon species (such as Chinook salmon) do not migrate to sea but mature to a small size in streams. All trout are spawned in freshwater, but may be exclusively nonanadromous or resident (that is, they spend their whole lives in freshwater), a mix of anadromous and resident, or virtually all anadromous. Each salmon and trout species is structured into discrete breeding populations because the adults return to their natal site to spawn (Quinn 2005). Therefore, the population, rather than the species, is the fundamental unit of conservation, and this is why DPSs of Pacific salmon and trout have been listed.

As a consequence of the complex population structure of Pacific salmon, some breeding populations can be highly endangered whereas other populations of the same species are abundant and able to sustain substantial exploitation from fisheries—for example, sockeye salmon in the Stanley Basin of Idaho vs those in Alaska's Bristol Bay (Hilborn et al. 2003; Gustafson et al. 2007). Pacific salmon and trout populations also vary considerably in life-history patterns, including the timing of a series of key events: the return migration by adults from the ocean to freshwater, the spawning season, the emergence of juveniles from gravel nests, the duration of residence in freshwater, and migration to sea (Quinn 2005). Therefore, depending on the species and populations in question, fish might be present in one river at vulnerable times of their lives and absent from another river at the same time of that year, and these variations in life-history traits could affect how salmonids are exposed to pesticides. For example, some juvenile Chinook salmon migrate from their natal streams to the ocean in their first summer of life whereas other juveniles of the same species spend a full year in the river system before migrating to sea (Taylor 1990; Healey 1991). The committee notes that the variation in spatial and temporal distribution of juvenile salmon residing in and migrating from river systems is further complicated by the substantial numbers of hatchery-produced juveniles, whose differences from wild fish in size, growth rate, and release timing can all affect migration patterns (Giorgi et al. 1997; Beckman et al. 1998).

Sublethal effects on sensory capacity, reaction, swimming ability, buoyancy control, or other aspects of performance might increase mortality. For example, chlorpyrifos, a common organophosphate insecticide, inhibited acetylcholinesterase in the brain and muscle of salmonids and affected spontaneous swimming and feeding behaviors of juvenile coho salmon in a concentration-dependent manner in the laboratory (Sandahl et al. 2005). Whether and to what degree sublethal effects affect survival in natural conditions is not clear. Laboratory exposure of cutthroat trout to carbaryl, an insecticide applied to oyster beds in some estuaries, affected swimming performance and predator avoidance (Labenia et al. 2007). It is certainly plausible (and perhaps even parsimonious) to conclude that there will be effects on survival in natural settings if environmental concentrations and exposure durations are comparable with those in the laboratory experiments, but the magnitude of the effects in relation to other sources of mortality is difficult to measure or model. Another complication in modeling the effects of pesticide exposure is that salmonids often prey on other salmonids (Duffy and Beauchamp 2008).

Moreover, if the population as a whole is stressed by factors that increase mortality over natural levels—such as water diversions that reduce flows, dams that alter sediment transport patterns, shoreline development in rivers or estuaries, or predation by nonnative species—the cumulative effects of the many stressors might be sufficient to put populations in jeopardy even though any single stressor, such as pesticide exposure, could have been sustained. Chemicals can also have indirect effects on individual organisms and the population. For example, most of the diet of juvenile salmon and trout in streams consists of insects, both larval stages of aquatic insects and terrestrial insects that fall on the stream surface (Nielsen 1992). Reductions in the prey base by pesticides might affect growth rate and life-history transitions that depend on growth (Mangel and Satterthwaite 1998) and have subtle but profound effects on fitness. Analogously, shifts in the insect community and changes in fish behavior associated with fine sediment in the stream bottom might reduce growth and survival of juvenile steelhead (Suttle et al. 2004).

Finally, the variation in life-history traits, between and even within species and subspecies, reinforces the importance of knowing the ecology of the particular species and population of concern for ecological risk assessment. It also highlights the difficulty of identifying a reliable surrogate species for testing and analysis, in particular a species whose life history is similar to that of the species of interest. For example, pink salmon generally migrate the short distance to the sea as soon as they emerge as free-swimming fry whereas juvenile Chinook salmon usually remain in freshwater for months to a year

(Continued)

> **BOX 4-1 Continued**
>
> and coho for more than a year. Pink salmon usually spawn within a few kilometers or tens of kilometers of the sea whereas Chinook salmon can migrate 1,500 km upstream or more to spawn, so their juveniles have to migrate the same distance to return to the sea. The different species also have different preferences for spawning substrate, stream sizes, and spawning seasons, all of which vary among their geographic distributions. Thus, the choice of a surrogate species for analysis and testing is challenging and complex at best. Even more challenging are the intraspecific variations in behavior, physiology, and distribution. For example, stream-type and ocean-type Chinook salmon differ in many attributes (Quinn 2005) that could affect exposure and vulnerability to pesticides. All the variation further emphasizes the need to assess the suitability of the surrogates and the applicability of the laboratory tests carefully when making decisions about likely effects of pesticides and other chemicals on listed species (Macneale et al. 2010).

Pesticides can indirectly affect a given species via effects on other species in the community. Indirect-effects analysis examines how a pesticide affects the habitat of a species. Because the indirect effects of pesticides on the species of concern can be favorable or unfavorable, it is more appropriately described as effects analysis than as hazard analysis. For example, the prey of the species of interest might be reduced in abundance or eliminated by the pesticide, perhaps because the prey is the target pest species or is affected along with the species of interest. Alternatively, populations of its predator or competitor might be reduced and the abundance of the species of interest thereby increased.

Because some indirect effects can be quantified, the committee recommends that they be incorporated into effects analysis. For example, for a situation in which food is the limiting factor and the major indirect effect is a 50% reduction in the food resource of the species of interest, the indirect effect can be incorporated into the population model by a 50% reduction in carrying capacity (maximum population size that can be supported by a specified area). In most cases, determining and quantifying such effects are more challenging and might require a conceptual model that incorporates the major components and linkages of the species' habitat that would respond to pesticide applications (see section "Effects Models" below). The modeling would entail an understanding of the ecology of all the species that might be at risk from pesticide exposure that live in the same area as and use resources similar to those of the listed species. There might be multiple nodes and links between affected species and the species in question, which might result in a fairly complex community dynamics model.

There are many candidate models and associated computer software for simulating community and ecosystem interactions (see, for example, Verhoef and Morin 2010). The primary hurdle in their use in decision-making applications is the large number of parameters that are poorly known, which results in substantial implicit uncertainty. Because of the uncertainty, it is important when using such modeling tools to strive to estimate component uncertainties quantitatively in a realistic and scientifically defensible way and to propagate all the component uncertainties through the community-level analysis formally and explicitly. Such methods as Bayesian networks and Monte Carlo approaches for quantifying uncertainty in analyses were discussed in Chapter 2. If quantitative information about community relationships is lacking, a qualitative modeling approach could be considered, such as signed digraphs, loop analysis, and matrix analysis (Puccia and Levins 1991). Those types of modeling can help to determine which variables should be included in a community or ecosystem model and can provide insight into which ones should be measured to provide the greatest reduction in uncertainty.

As in the different approaches used to evaluate sublethal effects, EPA and the Services appear to differ (on the basis of their responses to committee questions) in the extent to which they consider indirect effects. EPA (2012a, p. 22) stated that "if the best available biological information for a listed species does not establish a relationship between the affected taxa and the listed species, EPA believes that a no effect conclusion is warranted." That approach is logical, but *relationship* is not defined. FWS (2012, p. 5) stated that EPA does not consider potentially important "tertiary" effects and that "community-level effects are not considered." FWS (2012, p. 5) continued that EPA "only considers potential direct effects to those resources immediately relevant to the listed species." Likewise, NMFS (2012, p. 4) stated that "aspects such as prey dynamics (e.g., how quickly prey availability returns to background levels) and trophic consequences of herbicide applications are not considered" by EPA.

EPA uses a chemocentric approach to the assessment and begins with what is known about a chemical and its potential to affect various attributes of species' habitat. The Services take a species-centric approach and describe what is known about the life history of the species of concern, from which they infer the potential for pesticide-related effects. The different approaches seem to follow the same pattern as those used to evaluate sublethal effects, in which EPA takes a more quantitative approach and the Services a more qualitative approach. However, both quantitative information and qualitative information are necessary for comprehensive ecological assessments of the interactions of xenobiotic chemicals with the critical features of a species' habitat. Development of a species-specific conceptual model during the problem-formulation phase of the ecological risk assessment includes a specific enumeration of the important habitat components, which can then be addressed quantitatively or qualitatively—depending on the available information—during the effects analysis. The FIFRA Endangered Species Task Force has already begun to gather information on habitat and niche requirements of endangered species into an electronic database accessible to EPA and the Services (FESTF 2012).

Cumulative Effects

In the context of the ESA, cumulative effects are defined as "those effects of future State or private activities, not involving Federal activities that are reasonably certain to occur within the action area of the Federal action subject to consultation" (50 CFR 402.02). As is the case with indirect effects, that definition is not the common definition used by many ecologists who tend to use the definition promulgated by the Council on Environmental Quality (CEQ) under the National Environmental Policy Act (40 CFR Parts 1500-1508, 1978) in which a cumulative effect is "the incremental [effect] of [an] action when added to other past, present, and reasonably foreseeable future actions." In other words, cumulative effects are ones that "interact or accumulate over time and space, either through repetition or in combination with other effects" (NRC 2003, p. 2). However, the regulatory definition in 50 CFR 402.02 becomes much more like the CEQ definition if one incorporates the "environmental baseline," which includes past and present conditions. The committee could not determine a scientific basis for excluding other federal actions from the consideration of cumulative effects. Present and past federal actions *are* included in the environmental baseline. Therefore, in the following discussion, the committee's understanding of cumulative effects incorporates the environmental baseline. The committee notes that cumulative effects are related to aggregate effects—effects that result from exposure through multiple pathways. However, such effects would also be captured by considering or incorporating baseline conditions

Species live in variable environments and are constantly subjected to a variety of stressors. Some stressors, such as extreme weather, are stochastic (random and inherently unpredictable in magnitude and frequency) and might act on populations in a non-density-dependent fashion. In other words, the effects will be the same regardless of how many organisms are present. Other stressors, such as parasitism and predation, are more predictable and are density-dependent (they depend on the number of organisms present). Exposure to pesticides is one of many exogenous stressors that might influence the type and degree of response of species (Coors and De Meester 2008). Rohr et al. (2006) proposed using concepts in community ecology and evolutionary theory to provide insights about cumulative effects of pesticides and other anthropogenic or natural stressors. Their approach encompasses the use of direct and indirect effects of pesticide applications to assess the sensitivity of various communities and to identify which stressors will have the greatest effect.

The stressors that currently affect listed species are considered part of the environmental baseline conditions. Therefore, the interaction of existing stressors with the pesticides under consideration is within the purview of the Services and appropriately part of a BiOp. EPA, as the action agency, is responsible for providing the Services with any information that is known about how toxicity of a pesticide is modified by environmental factors (for example, effects of cold

Exposure 101

stress on pesticide toxicity). The responses to multiple stressors that are likely to have an effect (or have an increased effect) in the future are the cumulative effects. The committee has concluded that population models (see section "Effects Models at the Population Level" below) provide an objective, quantitative, and practical framework for incorporating baseline conditions and projected future cumulative effects into the ecological risk assessment in a way that is relevant to the requirements of the ESA. For example, a population model can represent the direct effects estimated from concentration-response relationships as reductions from baseline in survival and reproductive success and also can include effects on survival and reproduction of current and future habitat loss (as decreasing carrying capacity), habitat fragmentation (as changes in the spatial structure of the model), and climate change (for example, as increases in temporal variability of survival and fecundity to simulate the effect of an increase in frequency of extreme weather events). Such an approach will necessarily be chemocentric because the pesticide is the additive stress, but the approach also takes into account species-environment interactions and includes the effects of stressors other than the pesticide on a species.

In some cases, the pesticide being assessed has been in use for a long time, and the baseline population model already includes pesticide-induced reductions in survival and fecundity. Therefore, the calculated reductions in survival and fecundity are added to the baseline model's survival and fecundity (thus increasing their values) to obtain a model that simulates the dynamics of a population that is not exposed to the pesticide. The difference between the projections of that model and of the baseline model is an estimate of the degree to which current use and past use of the pesticide are contributing to the risks faced by a listed species or preventing its recovery. Thus, the risk assessor uses the information (risks with and without the pesticide) to inform the reregistration decision. The procedure described here does not require any more data than the case in which the baseline data are coming from populations that are not exposed to a pesticide.

EFFECTS MODELS

Effects models are used to characterize the effects of a pesticide at the individual level (effects on survival and reproduction) and at the population level (effects on population viability and recovery). EPA and NMFS use different models to evaluate the potential effects of a pesticide active ingredient on listed species and critical habitat. As described in its overview of ecological risk assessments for listed species (EPA 2004), EPA does not use effects models in its assessments. It assesses direct effects associated with different pesticide concentrations by using a risk-quotient (RQ) model that involves dividing an estimated exposure concentration by an effect concentration based on various prescribed toxicity tests and on published data. The derived RQ is compared with various levels of concern (LOCs) to determine whether a direct effect is likely. During its Step 2 assessments, EPA might also use direct-effect LOCs to draw inferences about the potential for indirect effects on listed species that rely on nonendangered organisms as critical food or shelter resources. The indirect-effects analysis also serves as the basis for analyzing potential effects on designated critical habitat. Population effects are addressed simply as an extension of individual effects; if survival or reproduction is affected, EPA assumes population-level consequences and makes a "likely to adversely affect" determination, which requires formal consultation with the Services (Step 3, see Figure 2-1). See Chapter 5 for further discussion of the RQ approach.

NMFS uses population models as one of several lines of evidence to address the question of population persistence explicitly. The BiOp on the effect of three pesticides on salmonids (NMFS 2008) served as an example of the NMFS modeling approach for the committee. In that BiOp, NMFS assessed risk by examining the overlap in the estimated environmental concentration (EEC) and effect concentration ranges, using a literature survey of effects observed in the field, and using a weight-of-evidence analysis for multiple lines of evidence applicable to a number of risk hypotheses. NMFS also evaluated potential effects of pesticides on populations with two models: a life-history population model that estimated changes in a population's rate of growth (lambda) on the basis of reduced individual survival after a 4-day exposure to acutely lethal concentrations and an individual-based growth and life-history population model that also estimated changes in lambda on the basis of reductions in growth of juveniles due to acetylcholinesterase inhibition and reduced prey abundances. That modeling was not done for a specific EEC but for a range of possible environmental concentrations that could be related to an EEC.

The committee was asked to consider the various approaches for evaluating pesticide effects, and it interpreted its specific task concerning models to be an assessment of modeling approaches at the individual and population levels, a clarification of the relationship between models at these two levels, and an evaluation of the major assumptions of the models. The following sections address those topics.

Effects Models at the Individual Level

All chemicals affect organisms through interactions at the cellular level—for example, binding to cell receptors and inducing or blocking normal responses, inhibiting or stimulating enzymes, causing cell death, or disrupting normal DNA replication. Some cellular changes result in measurable responses that might affect an organism's ability to survive and reproduce. Because organisms have redundant systems to maintain homeostasis and various mechanisms to detoxify and eliminate chemicals, there are exposures below which no organismal (individual) effects occur. However, individual organisms differ in their ability to tolerate chemical exposure, and this results in variability around the effects threshold. Variability throughout the toxic range is illustrated by a standard *con-*

Exposure 103

centration-response curve (also known as a dose-response or exposure-response curve), which is essentially a cumulative distribution function of the percentage of animals in a test population that exhibit a given response at each exposure concentration.

Superimposed on the interindividual (intraspecies) biological variability is variability from different sources, including interspecies variability, and uncertainty resulting from measurement imprecision and from extrapolation of experimental concentrations. Those types of uncertainty and approaches to incorporating them in individual-level models are discussed below in the section "Interspecies Extrapolations and Surrogate Species." The combined effects of those types of uncertainty can be expressed as confidence intervals around values on the concentration-response curve.

To evaluate potential effects on a species correctly, direct effects of pesticides on survival and reproduction must be estimated, and these estimates must correspond to the conditions expected in nature. The range of concentrations needs to include all plausible values that might result from the fate and transport models (see Chapter 3) for the populations that are being assessed. Because the values vary in space and time, the predicted effects on survival and reproduction also vary. The temporal and spatial variability in direct effects must then be incorporated into the population model to estimate population-level effects. An important source of uncertainty in this process is the measurement of direct effects on individual survival and reproduction under laboratory conditions, where demographic rates might be higher than in the natural environment of the species. Thus, the results of laboratory experiments need to be scaled to values expected in nature. There are two aspects of that scaling. First, the effects measured in the laboratory must be used to estimate the toxicant's effects in nature by taking into account the relative periods of exposure in the laboratory experiments (Pe) and in the wild (Pw). That step is not necessary if the two exposure periods are about the same. However, if there is a substantial difference, an adjustment might be necessary. For example, if the experimental mortality (Me) is measured over a 4-day period, but the exposure in the wild is estimated from exposure models to be, for example, 8 days, the overall mortality in the wild (Mw) might be higher than in the laboratory. How much higher depends on assumptions about how the pesticide affects individual organisms. An extreme assumption would be that all organisms that are highly responsive to the pesticide die in the first 4 days and that the mortality over 4-day and 8-day periods would be the same (Mw = Me). Another assumption could be that mortality in the wild is the same during each 4-day period. Thus, pesticide mortality in the wild would be calculated as $Mw = 1 - (1 - Me)^{Pw/Pe}$. Second, the estimated toxicant mortality must be combined with the natural mortality. For example, if pesticide mortality and natural mortality are independent, the survival rate in the natural environment of the species can be calculated as $(1 - Mw)(S)$, where S is the survival rate in nature without any pesticide effects. In some cases, the calculated mortality is expected to be in the baseline model because the pesticide has been in use and the model parameters are based on a population exposed to the pesticide (see the section "Cumulative Effects" above). In that case, the survival rate in the model (S) already includes the calculated pesticide mortality (Mw). To obtain a model that simulates the dynamics of a population that is not exposed to the pesticide, the survival rate would be calculated as S/(1 - Mw), again assuming that pesticide mortality and natural mortality are independent. Although the examples in this section are given only for mortality effects, similar calculations also need to be done for the reproductive component of the effects data.

The committee notes that the effects end point is often summarized as a single point on the concentration-response curve, such as the concentration that kills 50% of the test population (LC_{50}). However, for the purposes of population modeling as discussed below, the effects must be estimated at a range of concentrations that includes all values that the populations that are being assessed might plausibly experience. Therefore, the committee concludes that test results expressed only as threshold values or point estimates—for example, the no-observed-adverse-effect level, the lowest observed-adverse-effect level, or the LC_{50}—provide insufficient information for a population-level risk assessment.

Effects Models at the Population Level

The results of the effects model (the changes in survival and reproductive success as a function of pesticide exposure) are used in population models to assess effects on listed species. Population models are used to estimate population-level end points—such as population growth rate, probability of population survival (population viability), and probability of population recovery—on the basis of individual-level effects. Because the ESA is concerned with species or listed units within named species, the effects of pesticides must be expressed at the population and species levels. Therefore, the committee concludes that population models are necessary to quantify the effects of pesticides on populations of listed species.

The need for effects analyses to be conducted at the population level has been emphasized for at least the last 2 decades (see, for example, Cairns and Pratt 1993; Baird et al.1996; Ferson et al. 1996; Munns et al. 1997; Forbes and Calow 1999) and has been covered in several recent books (see, for example, Pastorok et al. 2002; Akçakaya et al. 2008; Barnthouse et al. 2007; Stark 2012). The applications of population models for effects analyses are too many to list comprehensively; some examples are Munns et al. (1997), Kuhn et al. (2000), Topping et al. (2005), Duchet et al. (2010), Willson et al. (2012).

Other types of models that have been used to assess ecological risks posed by pesticides and other toxic chemicals include models of individual organisms, bioenergetics models, and community and ecosystem models. As noted, the focus on population models in this report is necessitated by the specific requirement of predicting effects on listed species (for example, the risk of extinction) under the ESA. Other modeling types are appropriate for estimating other types

Exposure 105

of ecological effects; however, for calculating the probability of extinction or decline of a listed species, demographic population models are the most practical and relevant tools available.

Using a population model requires three inputs. Two of the inputs are the outputs of the exposure and effects models described previously. Effects models describe the change in population-model parameters (survival and reproduction) as a function of pesticide concentration, and exposure models provide estimates of pesticide concentration over time and space. The third input is demographic and life-history information, such as age at first reproduction, age-specific (or stage-specific) survival and fecundity rates over time and space in natural populations, and mechanisms and magnitude of density-dependent processes.

There is a large variety of population models, from deterministic, exponential models of a single population to stochastic, age-structured or stage-structured, spatially explicit metapopulation models with complex forms of density dependence (see introductions and reviews in Burgman et al. 1993; Akçakaya et al. 1999, 2008; Quinn and Deriso 1999; Caswell 2001; Morris and Doak 2002; Pastorok et al. 2002 for topics covered in the sections that follow). The appropriate models for purposes of pesticide-effects modeling are complex, species-specific models that incorporate all the relevant demographic parameters and spatial structure required to predict extinction risk. Some species, such as North American Pacific and Atlantic salmon, have been carefully studied and probably have sufficient data to assign values to parameters in such models. However, many listed species have been studied in only a cursory manner, and modelers have only enough information to characterize the life history of a group of species and are only able to use simple, generic, deterministic models that predict lambda, the finite rate of increase in the population. The committee concludes that in the absence of detailed demographic information, it is appropriate to use such models to characterize the baseline condition of a listed species, provided that the analyst incorporates estimates of uncertainty—for example, by using reasonable "high" and "low" demographic inputs—to bound the range of probable lambdas and includes a discussion in the final risk assessment about the magnitude of the uncertainty resulting from this lack of knowledge.

The sections that follow discuss important issues related to various components of population models that are especially relevant to assessing the risks posed by pesticide exposure.

Temporal Scale

The temporal scale of an assessment has two components: the time step of the model and the time horizon (duration) of the assessment. For most species in temperate ecosystems with generation times of 1 year or longer, an annual time step is appropriate. Except for the simplest models, whose main result consists of asymptotic measures of population performance (such as lambda), models that estimate population viability require specification of a time horizon. There is a tradeoff between relatively short time horizons, which allow more reliable projections but might not be relevant for the conservation of the species (because the goal is long-term existence of the species), and relatively long time horizons, which are more relevant but result in more uncertain projections of population viability. Even if the effect at the individual level occurs for only a few years, population-level effects might be observed longer because of changes in the age structure of the population. To account for such transitory effects, an assessment can use a time horizon of several generations of the species or the period during which a pesticide is expected to affect the population, whichever is longer.

Spatial Scale

The spatial scale of an assessment has two components: resolution and extent. For most population models, the spatial resolution should coincide with the typical sizes of the areas (or ranges of sizes) occupied by populations or subpopulations of the species. That might require a translation of the results of the exposure model to reduce the spatial resolution to a level that is appropriate for the species. In other words, the results of exposure modeling at very fine resolution (for example, 30-m grid cells for a species with a 1-ha home range and populations occupying areas of several square kilometers) can be translated into effects at the population level by calculating an overall reduction in survival and reproduction in each population on the basis of the average EEC to which the population will be exposed. The average EEC would be estimated with errors by the exposure model, and the errors would be incorporated by using joint probability distributions (see Chapter 5).

Ideally, the spatial extent of the models would include all areas in which a modeled species is exposed to the pesticide being evaluated. Both the spatial distribution of the species and the distribution of pesticide in the landscape might be heterogeneous. As a result, different populations of the species might be exposed to different concentrations of the pesticide, and even individual organisms in a population might have different exposures. In some cases, spatial variability of exposure can lead to source-sink dynamics in a metapopulation (Palmqvist and Forbes 2008).[1] That is, populations that are exposed to the pesticide might become sink populations[2] and thus deplete the populations that are not exposed; conversely, exposed populations might remain extant despite exposure because of dispersal from the unexposed populations in the same metapopulation (Spromberg and Johnson 2008). Accordingly, if there is dispersal between populations, exposure of one population can cause a reduction in an-

---

[1] A metapopulation is a set of populations of the same species in the same general geographic area that might exchange individual organisms through dispersal.

[2] A sink population has more deaths than births and remains extant only because there are more immigrants than emigrants.

other, unexposed population. Depending on the spatial separation of the areas, separate assessments can be performed for each area or a single assessment can be performed with a metapopulation model that represents each area as one or more populations.

The spatial variability of exposure would be estimated on the basis of spatially explicit projections of the exposure models, and the spatial variability in the species distribution would be based on the projections of a species-distribution model (an ecological-niche model or habitat-suitability model) that might be based on geospatial data (see the section "Characterization and Delineation of Habitat" in Chapter 3). The committee concludes that in the absence of spatial data, it is appropriate to use generic, single-population models with no spatial structure that include average exposure and environmental conditions expected in the exposed area of the species' range and to incorporate errors estimated with exposure modeling.

Temporal Variability

Variability (or stochasticity) refers to parameters of a population model that vary randomly, such as survival rates or fecundities in different age classes. Temporal variability means that models cannot predict the population size in the future precisely. Instead, they can project statistical distributions of future population sizes. The distributions are often used to calculate risks, such as risk of species extinction, risk of population extirpation, or risk of population decline to a predetermined level. Incorporating temporal variability results in a more realistic model that has more relevant end points, such as extinction risk. The committee concludes that population models that incorporate temporal variability and focus on probabilistic results are needed for assessing risks at the population level and that deterministic models are insufficient for this task. However, in the absence of such information, deterministic models with such end points as lambda (the finite rate of increase) can be used as the initial step of risk assessment. In such cases, every effort should be made to obtain the data necessary to estimate temporal variability, and the uncertainties in the end points reported should be clearly described in the assessment with the recognition that a deterministic baseline model might bias the assessment. Notwithstanding the use of a deterministic baseline model, uncertainties in the exposure analysis and the dose-response analysis should be incorporated into a risk assessment, for example, by using joint probability distributions (see Chapter 5).

Density Dependence

Density dependence (most commonly, the reduction in fecundity and survival that occurs as population size increases and that results from competition for food, breeding habitat, or other critical resources) is an important aspect of the dynamics of many populations and their responses to toxicants (Forbes et al.

2001, 2003). In the absence of data on effects of density on population growth and for screening-level assessments, it is reasonable to use density-independent models. Such models often use population growth rate as the main result, although if the models are stochastic, they can also be used to estimate population viability (the probability of population decline or extinction over a specified period). Although density-independent models make a number of assumptions and leave out important aspects of population dynamics, their results are more relevant for assessing pesticide effects on species than the results of models that assess pesticide effects only on individual organisms.

If there is evidence that survival or reproduction changes as a function of population density, it is important to incorporate density dependence into a model. That a species is rare or has been in decline does not necessarily mean that its dynamics are not density-dependent. For example, if the species has been declining because of habitat loss, its dynamics are probably density-dependent. In addition, species that have declined to very low abundances might be subject to depensation or inverse density dependence, which is the reduction in survival or fecundity that occurs at low density and accelerates the species' decline and which is commonly referred to as Allee effects (Courchamp et al. 2008).

Incorporating density dependence into a model of a population whose vital rates (survival or fecundity) might be affected by pesticide exposure presents challenges (Moe 2007). For example, the pesticide exposure might reduce the growth rate of the population by the same amount regardless of population size. Those conditions would make the density-dependence functions of baseline and effects models (population models with and without pesticide exposure) have the same shape (Figure 4-1A). In other cases, the pesticide effects on the growth rate of the population might be stronger in large populations (Figure 4-1B) and result in more-than-additive (synergistic) effects, or the pesticide effects might be stronger in small populations (Figure 4-1C) and result in less-than-additive (antagonistic) effects (see, for example, Forbes et al. 2001; Moe 2007). Thus, pesticide exposure might reduce the carrying capacity (or equilibrium population size) directly (by reducing survival and fecundity at all densities) or indirectly (by, for example, reducing abundance of species on which the species of interest preys). Whether the effect will be additive, synergistic, or antagonistic depends on several factors, including which life-history stages are affected by toxicity and density dependence (Forbes et al. 2001). The committee concludes that it is not accurate to assume that mortality due to pesticide exposure will be compensated for by density dependence; it is likely that such exposure will decrease the growth rate of a population at all densities and generally depress the population growth-density curve as depicted in Figure 4-1.

MIXTURES

Effects analysis requires knowledge or judgment of the adverse effects associated with individual chemicals or chemical combinations at concentrations

Exposure 109

predicted to occur in the exposure environment. The toxicity of a chemical mixture probably will not be known, and it is not feasible to measure the toxicity of all pesticide formulations, tank mixtures, and environmental mixtures. Therefore, combined effects must be predicted on the basis of models that reflect known principles of the combined toxic action of chemicals (El-Masri et al. 1997). This section discusses the state of the science of mixture toxicity, raises practical issues associated with mixture assessments, and provides a case study of the application of information in the context of assessing risks to listed species posed by pesticides.

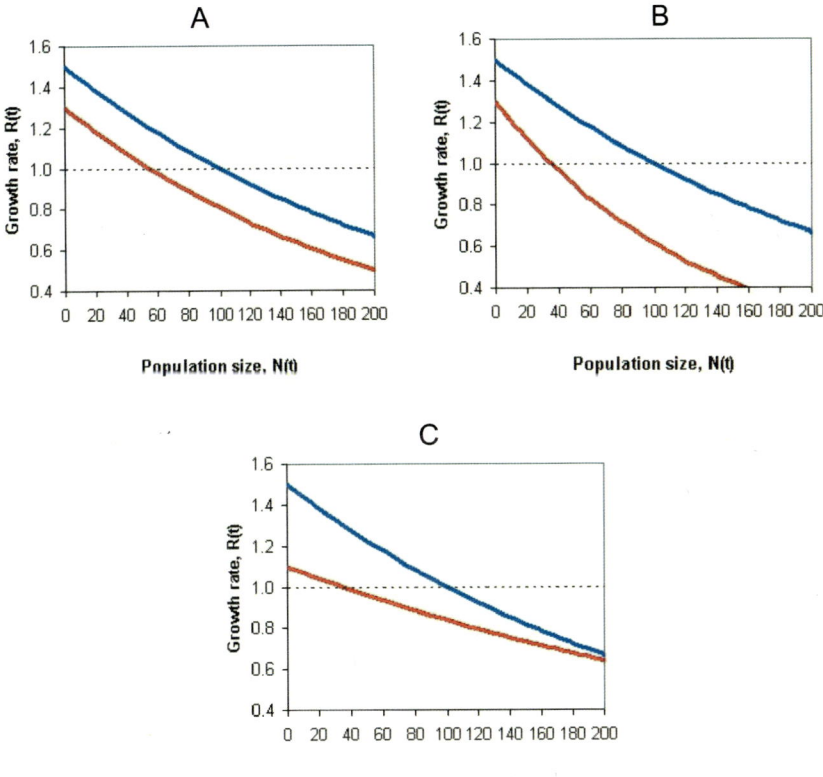

FIGURE 4-1 The effect of pesticide exposure on a density-dependence function. In all three graphs, the top curve shows the baseline model, and the bottom curve shows the effect model. Each curve shows the effect of density on population growth rate. Pesticide effects might decrease population growth equally at all densities (A), more at higher densities (B), or more at lower densities (C). Source: RAMAS 2011®. Reprinted with permission; copyright 2011, RAMAS®. RAMAS® is a registered trademark of Applied Biomathematics. See also Forbes et al. (2001) and Moe (2007).

Additivity and Interactions

The term *additivity* is used to designate forms of joint action in which the response to a mixture can be modeled on the basis of the expected responses to the mixture components in the absence of any toxic interactions. Two forms of additivity—concentration addition and response addition—are generally considered.[3] Concentration addition assumes that the components of the mixture act by the same mechanism[4] and that the components differ from each other only in their potency. Response addition assumes that the response to the mixture can be predicted on the basis of the expected responses to the individual components of the mixture. Toxic interactions are cases in which the joint toxic action of mixture constituents cannot be adequately described on the basis of additivity alone. Interactions are generally classified as synergistic (greater than additive) or antagonistic (less than additive). The frequency with which pesticide mixtures are found in surface waters is often cited as rationale for exploring the toxicity of pesticide mixtures (Scholz et al. 2006; Belden et al. 2007a; Laetz et al. 2009).

Concentration Addition

The central mechanistic assumption of concentration addition is that chemicals act by the same mechanism and differ from each other only in relative potency, with potency defined as the ratio of equitoxic doses. If the concentration of Chemical 1 associated with a given response rate is twice that of Chemical 2, Chemical 1 has half the potency of Chemical 2. Thus, relative potency can be used to convert an effective concentration of one chemical to a toxicologically equivalent concentration of another chemical.

Implicit in the application of concentration addition is the assumption that the slopes of the concentration-response curves for all mixture components are identical. The assumption of equal slopes follows directly from the assumption of functionally identical mechanisms of action. The slope of the concentration-response function is essentially a measure of the variability of individual tolerances in a population. Under the assumption that all chemicals in a mixture have the same mechanism of action, it follows that the distribution of individual tolerances and hence the shapes of the concentration-response curves will be the

---

[3]Concentration addition is also referred to in the literature as dose addition, simple similar action, or similar joint action. Response addition is also referred to in the literature as independent joint action or dissimilar joint action (see, for example, Bliss 1939; Finney 1971; EPA 2000). For consistency and simplicity, only the terms *concentration addition* and *response addition* are used in this discussion; it is recognized that *dose addition* is preferable to *concentration addition* when exposures are expressed as doses.

[4]*Mechanism* is defined in this context as the molecular interaction between a pesticide active ingredient and a biological target (for example, an enzyme or ion channel) that is responsible for the response being measured.

same for each chemical; hence, the slopes of the concentration-response functions of all the chemicals will be identical.

In practice, the slopes of the concentration-response functions will seldom be identical even for chemicals that have the same mechanism of action. Similarly, because of random variability, repeated bioassays of the same chemical on the same species by the same investigators will seldom have identical slopes. In such cases, methods are available for testing the significance of the differences between slopes and for constraining slopes to be parallel (Finney 1971). If the slopes of the concentration-response curves are identical (or can be constrained to be so without a significant lack of fit), the selection of the reference chemical for defining relative potency is incidental. That is, changing the reference chemical will change the relative potency values but will have no effect on the estimate of the concentration-response curve for the mixture.

In some cases, chemicals with the same mechanism of action at the receptor level can differ from each other in other ways (for example, differences in metabolic pathways) that can lead to differences in slopes in whole-animal studies. If the slopes of chemicals that act (or presumably act) similarly do differ, relative potency will vary with the magnitude of the response, and the application of concentration addition will be inappropriate.

Concentration addition is attractive because it is mathematically simple and is often viewed as a conservative assumption. As discussed below, concentration addition will typically predict a response rate that is equal to or higher than any form of response addition; it is conservative in this sense. Some groups have recommended concentration-addition as a general default method for mixture risk assessment, particularly for screening-level assessments (IPCS 2009; Kortenkamp et al. 2012). The EPA guidance for mixture risk assessment, however, recommends that concentration addition be applied only to groups of similarly acting chemicals (EPA 2000, p. 11). The committee concludes that the utility of concentration addition as a predictive and unbiased model for assessing joint action depends heavily on the underlying assumptions of concentration addition—similar mechanisms of action and parallel slopes. If those conditions are met, relative potency will be constant for all concentrations, so relative potency can be used to convert the concentration of one chemical into an equivalent concentration of another chemical. That conversion can be used to add concentrations correctly. If the underlying assumptions of concentration addition are violated, however, there is no reason to expect its application to be predictive. Application of concentration addition in those cases might lead to substantial errors that underestimate or overestimate the actual risk. Therefore, although the concentration-addition model has been demonstrated to predict the toxicity of pesticide active-ingredient mixtures more accurately when the pesticide active ingredients have the same mechanism of action (Belden et al. 2007a), caution should be exercised in using concentration-addition modeling as a default approach.

Response Addition

Response addition is a form of noninteractive joint action in which the response to a mixture is estimated on the basis of the responses (rather than the concentrations) that would be expected from the components of the mixture. Response addition does not assume that the components of a mixture act by the same or even a similar mechanism and does not assume anything about the slopes of the concentration-response curves. The slopes of the concentration-response curves for chemicals that have different mechanisms of action might or might not differ significantly. The only requirement is that the chemicals produce the same effect. In most practical applications of response addition, the end point is mortality; however, response addition can be applied to any quantal response. Response addition can take various forms, depending on assumptions about the correlation of tolerances in the population.

A review of the literature on pesticide-mixture toxicity revealed that the response-addition model provided somewhat more accurate predictions of toxicity than the concentration-addition model when the pesticide active ingredients had different mechanisms of action (Belden et al. 2007b). Response addition also has been shown to provide more accurate estimates of toxicity of mixtures that consist of dissimilarly acting pesticide and nonpesticide chemicals (Walter et al. 2002; Backhaus et al. 2004).

Synergy

Arguably, the greatest concern in evaluating hazards and risks to listed species posed by chemical mixtures that contain pesticides is whether constituents of the mixtures act to enhance the toxicity of the pesticide active ingredient. Indeed, pesticide synergists are often included in pesticide formulations (Jones 1998) and can enhance the toxicity of an active ingredient to nontarget organisms by a factor of 100 (Sahay and Agarwal 1997). The activity of some pesticide active ingredients also is substantially enhanced when they are administered in combination with other pesticides. Finally, chemicals to which coexposure occurs might increase the toxicity of a pesticide active ingredient by increasing its bioavailability or potency in the environment of the exposed organism.

Pesticide Formulation Synergists

Pesticide formulation synergists typically function by inhibiting cytochrome P450-mediated inactivation of the active ingredient (Hodgson and Levi 2001). They can enhance the effects of some pyrethroids, organophosphates, carbamates, and rotenone. Formulation synergists include bucarpolate, dietholate, iprobenfos, jiajizengxiaolin, MGK 264, octachlorodipropyl ether, piperonyl butoxide (PBO), piperonyl cyclonene, piprotal, propyl isome, sesamex, sesamolin, sulfoxide, and zengxiaoan. PBO is among the most potent and widely

Exposure

used formulation synergist (EPA 2005). Because formulation synergists are specifically used to increase the potency of pesticide active ingredients, they are most likely to produce the greatest enhancement of pesticide toxicity.

Toxicity evaluations that used the amphipod *Hyalella azteca* revealed that coexposure to PBO caused up to about a sevenfold increase in the toxicity of permethrin (Amweg et al. 2006). The synergistic potency of PBO increased as exposure concentration increased with a threshold concentration of 2.3 μg/L in water. The threshold concentration for synergy to occur stands in contrast to PBO surface-water concentrations, which are typically less than 80 ng/L even after direct application to surface water for mosquito abatement (Orlando et al. 2003, 2004; LeBlanc et al. 2004; Amweg et al. 2006). Given that *H. azteca* is considered sensitive to pyrethroids (Werner et al. 2010), that PBO is considered the most potent of formulation synergists, and that PBO concentrations in surface water after application tend to be below concentrations necessary to elicit synergism, there is a low probability that synergists associated with pesticide formulations enhance the toxicity of pesticide active ingredients. The greatest probability of synergistic effects might be when synergist-containing pesticide formulations are applied directly to aquatic systems or when there is direct contact between the formulation and a species.

Synergistic Interactions among Active Ingredients

As discussed in Chapter 3, pesticide active ingredients have the potential to coexist in tank mixtures or as environmental mixtures. In some cases, the toxicity of pesticide active-ingredient combinations has been shown to be greater than additive. The synergy has been exploited in recommended tank formulations to treat pests. With respect to nontarget species, the synergy has been recognized as a potential source of enhanced ecological threat. The following are examples of known synergistic interactions between pesticide active ingredients.

Organophosphates and Carbamates. Organophosphates and carbamates share a mechanism of action: inhibition of the enzyme acetylcholinesterase. Accordingly, the joint toxicity of organophosphates and carbamates should conform to a concentration-addition model. Indeed, the in vitro inhibition of acetylcholinesterase activity in salmon brains by combinations of organophosphates and carbamates showed that to be the case (Scholz et al. 2006). However, in vivo exposure of salmon to binary combinations of organophosphates, carbamates, or a combination of organophosphate and carbamate resulted in greater inhibition of brain acetylcholinesterase activity than would be predicted by concentration addition (Laetz et al. 2009). Serine esterases are important in the metabolic detoxification of organophosphates and carbamates (Cashman et al. 1996). Studies have shown that those esterases can be selectively inhibited by binding of one substrate, which results in increased toxicity of another because of its reduced detoxification (Murphy et al. 1959; Clement 1984).

Pyrethroids and Organophosphates. Studies in rodents (Ortiz et al. 1995) and insects (Martin et al. 2003) have shown that combinations of pyrethroids and organophosphates can synergize to produce greater than additive toxicity. Pyrethroids also are metabolized by serine esterases (Cashman et al. 1996), and it is reasonable to assume that various combinations of pyrethroids, organophosphates, and carbamates would have the potential to elicit greater than additive toxicity through the inhibition of serine esterases and perhaps other detoxification enzymes (Bielza et al. 2007).

Ergosterol Biosynthesis-Inhibiting Fungicides and Pyrethroids. Ergosterol biosynthesis-inhibiting (EBI) fungicides are potent inhibitors of some cytochrome P450 enzymes, and this inhibition is responsible for their mode of toxicity to fungi (Thompson 1996). Some EBI fungicides are imidazoles (for example, prochloraz and clotrimazole), triazoles (for example, propiconazole and tebuconazole), and morpholines (for example, fenpropimorph and aldimorph). Coexposure to some EBI fungicides and pyrethroids has been shown to result in greater than additive toxicity, presumably because of the inhibition of P450-mediated detoxification of the pyrethroids. The synergistic effect of EBI fungicides has been detected in a wide array of terrestrial and aquatic organisms (Norgaard and Cedergreen 2010; Bjergager et al. 2012) and reviewed in Cedergreen et al. (2006).

Synergy at High Laboratory Concentrations

Demonstrations of synergistic toxicity under controlled laboratory conditions often are performed at high chemical concentrations that are toxic even in the absence of synergy (see, for example, Anderson and Zhu 2004; Laetz et al. 2009). However, such synergy is of little use in identifying and quantifying synergy at low, environmentally relevant concentrations.

Many of the toxic mechanisms by which interactions might occur are saturable processes (such as rates of absorption, metabolism, and excretion), and many are governed by Michaelis-Menten or similar kinetics. In such processes, there are probably interaction thresholds—concentrations below which interactions are not likely to occur or, if they occur, will be minimal and probably not detectable (Figure 4-2). Toxic interaction thresholds have been described in terrestrial mammals (Dobrev et al. 2001; Yang and Dennison 2007; El-Masri 2010) and in aquatic organisms (Mu and LeBlanc 2004; Rider and LeBlanc 2005). Although the assessment of interaction thresholds is evolving, the current view, informed by empirical data, is that they are in the range of toxicity thresholds of the individual components of the mixtures (Yang and Dennison 2007). Similar observations were made much earlier and before the formal discussion of interaction thresholds (see, for example, Feron et al. 1995). Interaction thresholds make sense in the context of the underlying kinetics and might be useful in assessing whether concerns about potential toxic interactions are important in exposures to specific mixtures.

Exposure 115

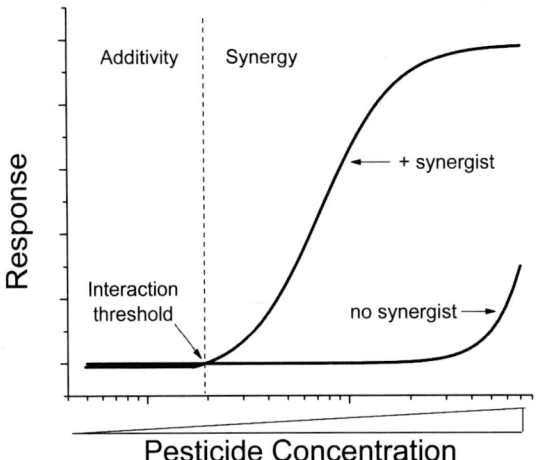

FIGURE 4-2 Concentration-response curve of a chemical in the presence and absence of a synergist. Toxicity of the chemicals is consistent with additivity below the interaction threshold and with synergy above the interaction threshold.

The existence of interaction thresholds does not necessarily reduce the probability of additive responses to mixtures of chemicals in which each chemical is below its own toxic threshold. In such cases, an additivity model might suggest the potential for a toxic response. The concept of interaction thresholds simply indicates that the probability of toxic interactions (as opposed to some form of additive joint action) is reduced if the total exposure does not exceed a threshold based on an assumption of additivity. A difficulty in the application of the concept of interaction thresholds is quantification of the threshold, which is difficult in the risk assessment of single chemicals and mixtures.

Uncertainty Factor to Account for Synergism

A specific charge to the committee was to "consider the selection and use of uncertainty factors to account for the lack of data on…synergy [and] additivity." The committee concludes that there is no scientific basis for applying a generic uncertainty factor under the presumption of a synergistic interaction. Doing so would introduce a bias into the risk assessment.

As an alternative to generic uncertainty factors, EPA's supplemental guidance for mixture risk assessment outlines a weight-of-evidence approach for incorporating quantitative consideration of interactions by using numerical binary weight-of-evidence scores that are based largely on qualitative information on potential interactions and any information on the magnitude of interactions of mixture components (EPA 2000, Section 4.3.1.1, pp. 90-103). That method was developed in the context of human-health risk assessment, and no examples of

Antagonism

Antagonism occurs when components of a mixture interact in a manner that results in toxicity that is less than would be predicted under an assumption of additivity. Antagonistic interactions that are most likely to affect pesticide toxicity occur when two components of a mixture are competing for the same target site of toxicity and the less toxic component competitively displaces the more toxic component or when a component of the mixture inhibits the metabolic conversion of a pesticide to a more toxic derivative. The former situation has been well described with binary combinations of pesticide active ingredients that share a mechanism of action. For example, exposure of the Asian catfish *Mystus vittatus* to the acetylcholinesterase inhibitors dichlorvos and thiotox or dichlorvos and carbofuran resulted in less toxicity than would be predicted on the basis of additivity (Verma et al. 1980). In both cases, exposure to the less toxic component at the maximum tolerated concentration reduced the toxicity of the more toxic constituent, presumably because of competitive displacement of the more toxic component from the target enzyme. Similar antagonistic effects have been observed with pairs of pyrethroid insecticides (Schleier and Peterson 2012).

Although PBO is typically used as a pesticide synergist by inhibiting cytochrome P450 activity, it and other cytochrome P450 inhibitors can decrease the toxicity of pesticides that are enzymatically converted to more toxic oxidative metabolites. For example, malathion and parathion are metabolically converted to their more toxic oxon derivatives by the actions of cytochrome P450s. Exposure of daphnids to either insecticide and PBO significantly reduced toxicity of the insecticides (Rider and LeBlanc 2005). Similar antagonism was observed with PBO and chlorpyrifos, which also is metabolically converted to the more toxic oxon derivative (Ankley and Collyard 1995; El-Merhibi et al. 2004).

The committee emphasizes that for a mixture component to antagonize (reduce) the toxicity of a pesticide active ingredient, the pesticide active ingredient must be present at a concentration that actually elicits toxicity. Given that circumstance, the committee concludes that ecological risk assessment should focus on the pesticide active ingredient alone and avoid the added uncertainties associated with estimating the reduction in risk due to the presence of an antagonist.

Complex Mixtures

Many environmental mixtures are highly complex, that is, contain a large number of components. That can complicate the exposure and effects analyses of the mixtures. As discussed at some length in EPA's supplemental guidance

Exposure

on mixture risk assessment (EPA 2000), confidence in the application of component-based methods diminishes as the number of components of a mixture increases. For some highly complex mixtures, such as petroleum distillates and surfactants, component-based methods might be impossible because the components are not well characterized and toxicity data on them are lacking.

For complex mixtures, the preferred assessment method is to use data on the whole mixture, termed the *mixture of concern*. The toxicity of the mixture is assessed with a bioassay. A problem with the mixture-of-concern approach, however, is that the composition of many complex mixtures is variable. Often, toxicity data are not available on the mixture of concern but are available on similar mixtures. Addressing those types of situations involves an assessment of *sufficient similarity*, that is, determining which, if any, of the mixtures on which data are available can be used to assess the toxicity of the mixture of concern reliably. EPA (2000, p. 38) offers only brief and general guidance on assessing sufficient similarity:

> In determining whether a mixture is sufficiently similar, consideration should be given to any available information on the components that differ or are contained in markedly different proportions from the mixture of concern. In addition, if information exists on differences in environmental fate, uptake and pharmacokinetics, bioavailability, or toxicological effects for either of these mixtures or their components, it should be considered in deciding on a risk assessment approach.

EPA (2000) also notes that the *comparative-potency method* might be useful in assessing toxicity of complex mixtures. Essentially, that method involves using toxicity data on complex mixtures for which two toxicity end points—for example, an LC_{50} and a reproductive no-observed-effect concentration (NOEC)—are known to estimate a toxicity value (say, a reproductive NOEC) for a mixture for which only the other end point (an LC_{50}) is known.

Simply because a mixture is complex does not indicate that the joint action of the mixture is complex. It is highly relevant to examine the frequency at which environmentally relevant chemical mixtures elicit cumulative toxicity. Olmstead and LeBlanc (2005a) evaluated the toxicity of a mixture of nine chemicals, including pesticides, at the median concentration in US surface waters as measured by Kolpin et al. (2002) and dilutions and fortifications thereof. The mixture elicited concentration-dependent toxicity at exposure concentrations between the median and 10 times the median concentrations of the chemicals, and the investigators were able to predict the toxicity of the mixture accurately with a model that combined concentration addition and response addition. However, further analyses revealed that the toxicity of the mixture could be explained largely by a single constituent, chlorpyrifos. The experiment was repeated without chlorpyrifos in the mixture. Toxicity was not eliminated by the removal of chlorpyrifos; rather, the remaining chemical mixture elicited toxicity at a higher concentration. Again, analyses of the responses to the individual chemicals in the mixtures revealed that toxicity of the new chemical mixture was due to the actions of a single constituent in the mixture, diazinon. Thus, the chemical mixtures, at environmentally relevant constituent ratios, produced toxicity, but toxicity could be attributed primarily to a single constituent.

Adam et al. (2009) evaluated the aggregate toxicity to the amphipod *Gammarus pulex* of four pesticides—propiconazole, tebuconazole, 3-iodo-2-propinyl butyl carbamate, and cypermethrin—at concentration ratios typically found in commercial formulations. Toxicity of the mixture could be explained by the concentration of cypermethrin. Studies of environmental samples that contain chemical mixtures have typically shown that toxicity of the mixture can be attributed to one or a few constituents (Amweg et al. 2006; Belden et al. 2007a).

Although mixtures of pesticides and other chemicals clearly exist in the environment, the presence of a mixture does not necessarily imply toxicity. Furthermore, on the basis of mixture modeling theory discussed above, the presence of dissimilarly acting chemicals, each present at a concentration that elicits no toxicity, would not be predicted to elicit toxicity in a mixture in the absence of synergy. The presence of chemicals that have the same mechanism of action, each at a concentration that elicits no toxicity, would be predicted to elicit toxicity only if the combined, potency-normalized concentrations of the chemicals exceed the threshold concentration for a response.

Practical Issues in Assessing Effects of Pesticide Mixtures

Assessment of the effects of pesticide mixtures is associated with many practical aspects. This section first describes the positions taken and approaches used by EPA and the Services, then discusses various issues associated with pesticide formulations, and finally provides a perspective on the magnitude of interactions.

Agency Positions and Approaches

Approaches to addressing risks associated with exposures to mixtures are clearly a major source of disagreement between EPA and the Services, although the discord is not based on any fundamental disagreement about methods. The BiOps prepared by NMFS (2008, 2009, 2010, 2011) suggest that general guidelines used by EPA have been adopted by NMFS inasmuch as they reference EPA methods for ecological risk assessment (EPA 1998a, 2004). In specific and quantitative considerations of mixture exposures (for example, NMFS 2010, p. 465ff), the Services adopt concentration addition for mixtures of similarly acting pesticides, and this approach is consistent with the methods recommended by EPA (EPA 1986, 1989, 1998a, 2000, 2002).

Although EPA and the Services appear to accept the same basic methods in mixture risk assessment, their implementations of the methods differ substan-

Exposure 119

tially. Each BiOp developed by NMFS expresses substantial concern for mixture exposures and the potential for synergistic effects. As discussed further below, similar concerns are also expressed by FWS. Although EPA guidelines (EPA 1998a, 2004) certainly recognize and appreciate the potential importance of exposure to mixtures, ecological risk assessments prepared by EPA focus on single active ingredients in the generic risk assessments (EPA 2004). The Services, in contrast, note a need to address exposures to all active ingredients *and* inerts that might affect populations of species (for example, FWS 2009).

Although diametrically opposed, the positions of the Services and EPA both have merit. EPA may elect to look at only single agents in most of its risk assessments, but agency-wide guidelines for mixture risk assessment, particularly EPA's *Supplementary Guidance for Conducting Health Risk Assessment of Chemical Mixtures* (EPA 2000), provide quantitative approaches for incorporating any available information on potential chemical interactions into a risk assessment. The approaches, however, are extremely labor-intensive, are accompanied by substantial uncertainties that are not readily quantified, and are best suited to site-specific analyses in which exposures to specific chemicals can be estimated with confidence. Thus, EPA takes the position that practical and unavoidable limitations in resources and data preclude a detailed quantitative assessment of chemical interactions, and the Services take the position that a quantitative assessment of interactions should be done. The committee concludes that quantitative assessment of chemical joint action is warranted if adequate data are available on the exposures to and toxicities of the chemicals. Approaches for such analyses are detailed further in the final section of this chapter, "Conclusions and Recommendations."

Formulation Toxicity

Several practical issues arise in the determination and use of data on pesticide formulations. There is the issue of availability of toxicity data on the inerts and the formulations themselves. There are also issues of the applicability of formulation data, particularly in considering long-term or chronic effects, and the applicability of formulation data that are extrapolated from other formulations. Those issues and others are discussed below.

Toxicity Data on Inerts

As discussed by Levine (1996), the original testing requirements for inerts in pesticide formulations were developed by the Food and Drug Administration, and these requirements were less rigorous than the testing requirements for the pesticides (that is, the active ingredients). In 2006, the EPA Inert Ingredient Assessment Branch completed a series of inert-ingredient tolerance-reassessment decision documents; however, documents covering all approved inerts do not appear to be available (EPA 2012b). One explanation is that tolerances are developed only for chemicals that are approved for use on food crops. List 1 inerts (toxic inerts) cannot be used on food crops and therefore do not have tolerance-reassessment documents. EPA has prepared guidance documents for companies involved in the development of new food-use inerts (EPA 20120c), new non-food-use inerts (EPA 2012d), and low-risk polymer inerts (EPA 2012e); additional guidance is provided in EPA's Pesticide Registration Manual (EPA 2012f). Although the documents do not contain an explicit list of required tests, they suggest that EPA could require the same types of tests as are required for active ingredients. With the exception of the guidance document on low-risk polymers, all the inerts guidance documents refer to the Office of Prevention, Pesticides, and Toxic Substances Harmonized Test Guidelines, which are used in the registration of pesticide active ingredients (EPA 2013). The committee found neither a specific list nor examples of the tests required for a new inert under EPA's guidance documents for developing new pesticide inerts.

Related to the availability of toxicity data and testing requirements for pesticide inerts is the determination of whether additional data are needed on them. EPA noted that it will often rely on acute toxicity studies of formulations in mammals to determine whether acute toxicity studies of formulations should be required in fish and invertebrate species (E. Odenkirchen, EPA, personal commun., April 4, 2012). Specifically, the agency referred to a series of standard studies often referred to collectively as the mammalian six-pack: acute oral toxicity (EPA 1998b), acute dermal toxicity (EPA 1996), acute inhalation toxicity (EPA 1998c), acute eye irritation (EPA 1998d), acute dermal irritation (EPA 1998e), and skin sensitization (EPA 2003). The extent to which those acute studies in mammals will be reliable in assessing the potential of inerts to enhance toxicity of pesticide active ingredients to fish and aquatic invertebrates is not clear. The committee identified no explicit analyses that support the use of mammalian six-pack studies of active ingredients and formulations to assess the potential of inerts to enhance toxicity to fish and aquatic invertebrates.

Toxicity Data on Formulations

Pesticide formulations are not tested as extensively as active ingredients. However, if data are available on both an active ingredient and a formulation, the contribution of inerts to the toxicity of the formulation can be at least crudely assessed. All pesticide formulations must identify the percentages of their active ingredients. If $\pi_{a.i.}$ is the proportion of an active ingredient in a formulation and ζ_F is the toxicity value for the formulation, such as the LC_{50}, the toxicity value of the active ingredient ($\zeta_{a.i.}$) is the product of those two terms. $\zeta_{a.i.}$ can then be compared with experimentally determined toxicity values of the active ingredient alone. If, for example, the LC_{50} of the active ingredient as part of the formulation is substantially lower than the LC_{50} of the active ingredient alone, it suggests that some components of the formulation might be contributing substantially to the toxicity of the formulation. Conversely, if the LC_{50} of the active ingredient as part of the formulation is substantially higher than the measured LC_{50} of the active ingredient alone, it suggests that some components of the formulation might be reducing the toxicity of the active ingredient. That type of comparison often yields the only type of quantitative information that can be used to assess the toxicological importance of inerts in a formulated product.

Chronic toxicity studies are not generally conducted on pesticide formulations and are not available on most inerts. In cases in which a chronic toxicity value is available for an active ingredient ($Ch_{a.i.}$) and acute toxicity values are available for the active ingredient alone ($Ac_{a.i.}$) and the active ingredient as part of the formulation (Ac_{Form}), a chronic toxicity value for the formulation (Ch_{Form}) could be estimated by using the following formula:

$$Ch_{Form} = \frac{Ac_{Form}}{Ac_{a.i.}} Ch_{a.i.} \quad \text{(Eq. 1)}$$

where all toxicity values are expressed in units of active ingredient.

The above approach raises important issues. A major assumption is that acute toxic potency and chronic toxic potency will be the same or at least closely related. Acute toxicity end points, such as mortality, will often bear little relationship to chronic end points, such as growth and reproductive capacity, and there will be little or no basis for assuming that the underlying mechanisms of action in producing acute and chronic effects are comparable. The approach becomes more palatable when the same mechanism-based end point is used for acute and chronic toxicity evaluations. For example, cholinesterase inhibition might be considered for assessing responses to acute and chronic exposures to an organophosphate or carbamate insecticide. Again, the major impediment to the use of the approach is the general lack of mechanism-based data on effects after chronic exposure.

Environmental Partitioning and Applicability of Formulation Studies

A major reservation in the use of formulation studies concerns environmental transport. Unless an active ingredient and inerts in a formulation have similar chemical-fate and environmental-fate properties—this is seldom the case—the active ingredient and the other components of the formulation will partition at different rates in various environmental compartments (see Chapter 3). If the partitioning is substantial, as is often the case, there is little rationale for asserting that differences in acute toxic potency will have any relationship to differences in effects of chronic exposures, a relationship that is assumed in the approach described above.

Considerations of environmental partitioning also affect the usefulness of chronic studies of formulated products. Although longer-term studies can be conducted on pesticide formulations, such studies typically involve designs in which partitioning processes do not occur. The studies can provide useful information on the long-term effects of a formulation as it is applied, but these exposures probably have little relevance to exposures that occur in the environment as environmental partitioning occurs.

In some cases—for example, the Roundup formulations of glyphosate (EPA 2008)—detailed information is available on the toxicity of an active ingredient, the toxicity of specific inerts, and the toxicity of a formulation. In such cases, more detailed assessments can be conducted on the basis of analyses of toxic interactions or available data on the mixture of concern.

Data-Bridging

Another problem that arises in using formulation studies to assess the toxicity of inerts in pesticide formulations concerns "data-bridging." Although EPA generally requires at least acute toxicity data on pesticide formulations, it will often allow toxicity studies on one formulation to support the registration of another. That general approach is sometimes referred to as bridging registration (EPA 2012f). Data-bridging is used in the United States (EPA 2002, 2012f) and member nations of the Organisation for Economic Co-operation and Development (OECD 2001). Although data-bridging is motivated by economic factors (reducing the costs associated with pesticide registration) and ethical factors (reducing the number of animals that must be used in toxicity studies), the process of data-bridging must be supported by similarities between the two formulations to be scientifically credible.

The committee concurs that data-bridging is sensible if two formulations are identical (the same formulation marketed under different names). If two formulations are substantially different, however, formulation-specific data are required. Recently, EPA (2012g) released relatively detailed guidance for waiving mammalian acute toxicity studies. Although the document is titled *Guidance for Waiving or Bridging of Mammalian Acute Toxicity Tests for Pesticides and Pesticide Products*, the main focus concerns the criteria for waiving acute toxicity studies rather than for bridging data among formulations. The OECD guidelines (OECD 2001, p. 89ff) articulate bridging principles that are generally consistent with EPA's approach to assessing sufficient similarity. However, formulation-bridging is not transparent. In the absence of information, uncertainty can lead to assumptions that are not justified.

Foreign Formulations

Information is sometimes available on the toxicity of pesticide formulations used outside the United States. If comparable information is available on the US and foreign formulations, the information on the US formulation should take precedence because it will be the most applicable. If information is not available on the US formulation, the relevance and utility of the information on the foreign formulation will be difficult to assess. For example, EPA's ecological risk assessment of glyphosate (EPA 2008) concludes that the probability that glyphosate will affect reproduction of mammals and birds is minimal, and similar conclusions have been drawn in several other risk assessments of glyphosate (WHO 1994; Giesy et al. 2000; Williams et al. 2000). One study in the South American literature, however, indicates that exposure to a South American formulation of Roundup, which contains glyphosate and proprietary surfactants, reduces testosterone concentrations in rats (Dallegrave et al. 2007). Similar reductions were observed in mallards after exposure to a South American formulation of Roundup (Oliveira et al. 2007). As part of the public comments on the registration review of glyphosate, it has been suggested that EPA use the data on the South American formulations to assess the risks to birds and mammals associated with exposures to US formulations (BeyondPesticides 2009); that is, the information on the foreign formulations should essentially be bridged to US formulations. Although pesticide manufacturers are required to disclose information on the composition of pesticide formulations registered in the United States to EPA, the requirement does not extend to pesticide formulations used only outside the United States. Therefore, EPA and the Services probably do not have access to information that would be useful in assessing the similarities or dissimilarities between US and foreign formulations. In the absence of that information, the merits of including data from studies of foreign formulations cannot be determined.

Magnitude of Interactions

EPA seldom addresses interactions quantitatively in ecological risk assessments of pesticides. In three BiOps, NMFS (2008, 2009, 2010) does address information on joint action and relies primarily on the publication by Laetz et al. (2009), which assayed acetylcholinesterase inhibition in Pacific salmon (*Oncorhynchus kisutch*) for all binary combinations of three organophosphates (diazinon, malathion, and chlorpyrifos) and two carbamates (carbaryl and carbofuran). The study by Laetz et al. (2009) notes interactions that range from additivity to synergism with an increasing prevalence of synergism as the exposures increased. The latter observation is consistent with the concept of interaction thresholds described above. The BiOps, however, do not attempt to use the information from Laetz et al. (2009) to adjust estimates of expected biological responses to mixtures quantitatively. NMFS (2009, p.266) noted that "we are unable to create a predictive model of synergistic toxicity as dose response relationships with multiple ratios of pesticides are not available and the mechanism remains to be determined." Other BiOps contain similar language and express concerns about mixtures.

EPA's agency-wide supplementary guidance for mixture risk assessment (EPA 2000) discusses methods for quantitatively addressing toxic interactions, but the methods are cumbersome to apply, and experiments that would be useful in assessing the predictive value of the methods seem not to have been conducted. The guidance document, however, reviews studies of the acute toxicity of all possible binary combinations of more than 50 industrial chemicals in rats (Smyth et al. 1969) and notes that deviations from the assumption of additivity span a factor of about 5 (expected to observed ratios of LD_{50} values of binary mixtures range from about 0.2 to 5). More recently, Boobis et al. (2011) searched publications on mammalian toxicology from 1990 to 2008 for studies reporting synergy at low sublethal doses. They discerned a maximum magnitude of synergy of about 3.5. The studies suggest that chemical interactions can modify toxicity, typically by less than a factor of 10. Exceptions do exist. PBO was shown to decrease the toxicity of malathion to daphnids (*Daphnia magna*) by as much as a factor of 100 (Rider and LeBlanc 2005), and the pesticide synergist *N*-octyl bicycloheptene dicarboximide increased the toxicity of deltamethrin to a snail (*Lymnaea acuminate*) by as much as a factor of 300 (Sahay and Agarwal 1997).

A Case Study: Assessing Pesticide-Containing Mixtures

Conventional approaches to assessing the risks posed by exposure to chemical mixtures first determine EECs of the mixture components and then estimate the hazard associated with those exposures. The approach is most appropriate for estimating whether the margin of safety for exposed listed species is sufficient. Table 4-1 provides a hypothetical dataset to exemplify how hazard from exposure to a pesticide in combination with other chemicals in the environment could be assessed by using established approaches discussed in this report. The listed species of concern in this exercise is sockeye salmon (*Oncorhynchus nerka*), and the pesticide of concern is cypermethrin, a pyrethroid. The cypermethrin is in a formulation that also contains the synergist PBO. The formulation will be added to a tank mixture that contains another pyrethroid insecticide, deltamethrin; a surfactant, Polysorbate-20; and a stabilizer, Epoxisoy. Contents of the tank mixture will be used for insect-pest control according to label specifications. And, several chemicals— nonylphenol, ethinyl estradiol, caffeine, acetaminophen, PBO, and cypermethrin—are known or predicted to be present in the exposure environment at measurable concentrations. Thus, the salmon might be exposed simultaneously to nine chemicals that originate from various sources. At issue is whether exposure to the chemical mixture poses a risk to the salmon.

Columns 2, 3, and 4 in Table 4-1 depict the sources of the chemicals and their concentrations in the different sources. Exposure of salmon to the mixture constituents can be estimated from those data. For example, cypermethrin is present in the formulation at a concentration of 0.05%. Once in the tank mixture, cypermethrin is diluted to a concentration of 250 mg/L (ppm), and the predicted concentration in the environment is 0.01 μg/L (ppb). The concentration of cypermethrin in the formulation can be derived from the label, the concentration in the tank mixture is calculated on the basis of the dilution to which the formulation is subjected, and the environmental concentration would be determined from modeling using, for example, PRZM2/EXAMS II.

Column 5 of Table 4-1 presents the final environmental exposure chemical concentrations, which are the sums of the concentrations of the individual chemicals in Column 4 (environmental exposure concentration). Most of the values are the same as presented in the previous column unless the chemical appears twice. For example, cypermethrin was present in the formulation to be applied with a residual amount (0.005 ppb) already present in the environment; thus, the final concentration of cypermethrin in the exposure environment is the sum of the two concentrations. The exposure analysis concerning the mixture of chemicals present in the environment is arguably the most challenging aspect of the mixture risk assessment owing to the high degree of uncertainty about the identity of chemicals and their environmental concentrations. Column 6 of Table 4-1 identifies chemicals expected to elicit toxicity by the same mechanism of action: chemicals that act similarly are assigned the same letter. Thus, cypermethrin and deltamethrin are both expected to elicit toxicity through the disruption of axonal sodium channels and are both assigned the letter a. Nonylphenol and polysorbate 20 have the ability to mobilize the pesticide and are identified with the letter b. Nonylphenol and ethinyl estradiol elicit estrogenic activity and are therefore identified with the letter c.

It is necessary at this stage to identify the adverse response of the listed species that is deemed most relevant to the pesticide of concern. In this exercise, disruption of axonal sodium channels is assumed to be the mechanism of action, and immobilization and loss of equilibrium are identified as the relevant responses because these sublethal responses are considered indicative of impending lethality or reproductive impairment. The other eight chemicals are considered only in their potential capacity to modify that response to cypermethrin. Thus, reproductive impairment associated with the combined estrogenicity of nonylphenol and ethinyl estradiol would not be considered relevant to the assessment of cypermethrin and would not be integrated into the toxicity assessment. If one or more components of the environmental mixture were predicted to elicit toxicity independently of cypermethrin, risk assessments of those components might be warranted.

Column 7 (K values) of Table 4-1 identifies chemicals that have the potential to enhance the toxicity of cypermethrin and the similarly acting chemical deltamethrin in a nonadditive manner. That would include synergists, PBO in this exercise. The modifying effect of the synergist PBO on the response to cypermethrin and deltamethrin is defined by a coefficient of interaction (K) (Mu and LeBlanc 2004; Rider and LeBlanc 2005; TenBrook et al. 2010), which can be viewed as a special case of the coefficient of synergism (κ) as described by Finney (1942). K values are typically determined experimentally by assessing the effect of increasing concentrations of a synergist on a specific response to a pesticide—such as an estimate of the effective concentration at which 50% of

TABLE 4-1 Example Dataset Used to Assess Exposure to and Effects of Cypermethrin, in Mixture with Several Other Chemicals, on the Sockeye Salmon, *Oncorhynchus nerka*

1	2	3	4	5	6	7	8	9
Chemical	Formulation (%)	Tank (ppm)	Environment (ppb)	Final Exposure Concentration (ppb)	Similar-Acting[a]	K Values	EC_{50} (ppb)	Slope
Formulation[b]								
Cypermethrin	0.05	250	0.01	0.015	a		5.0	14
PBO	1.0	5,000	0.2	0.205		Cyper: 1.0[c] Delta: 1.0[c]	10,000	5.0
Tank								
Deltamethrin		100	0.004	0.004	a		3.1	15
Polysorbate 20		10,000	0.40	0.40	b		25,000	2.3
Epoxisoy		5,000	0.20	0.20			75,000	3.3
Environment								
Nonylphenol			0.001	0.001	b,c		1,500	5.7
Ethinyl estradiol			0.05	0.05	c		1,400	6.1
Caffeine			500	500			33,000	3.9
Acetaminophen			0.01	0.01			100	6.3
PBO			0.005					
Cypermethrin			0.005					

[a]Chemicals that act similarly are assigned the same letter.
[b]Formulations can be evaluated for the potential contribution of inerts as described in the section "Formulation Toxicity."
[c]K values of 1.0 were assigned to the synergist PBO because the environmental exposure concentration of PBO in this exercise was considered to be below the interaction threshold.

the population exhibits a defined response (an EC_{50})—in a surrogate species (Figure 4-3). Thus a K value of 2.0 would indicate that the corresponding concentration of synergist in the exposure environment would be expected to increase the toxicity of a pesticide by a factor of two (Figure 4-3). Alternatively, K values can be estimated on the basis of the available literature on the modifying effect of a synergist on the toxicity of a pesticide. K values also can be derived for factors that affect the environmental availability of a chemical (for example, enhanced mobility or dissolution because of a surfactant). However, those modifying effects would be more appropriately addressed when modeling environmental exposures.

Columns 8 (EC_{50}) and 9 (Slope) of Table 4-1 define the toxicity of each chemical with respect to the response of concern (immobilization or loss of equilibrium in this exercise). The concentration-response curve is used to define the EC_{50} and the slope (or power) of the relationship. Those values are required for later mixture modeling. Typically, the data are determined experimentally by using a surrogate species after an appropriate exposure duration. EC_{50} data are often available from the literature, and EC_{50} and slope values are required for FIFRA guideline studies.

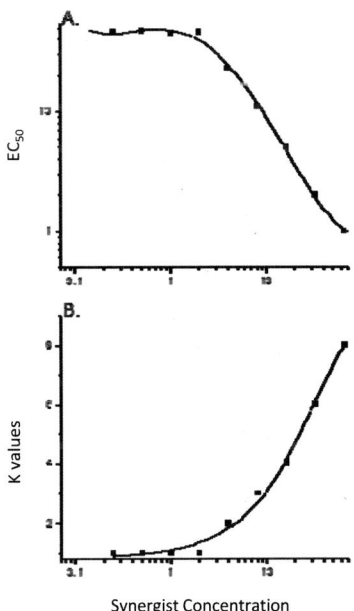

FIGURE 4-3 Example derivations used to determine K values. (A) Measured effect of the synergist on the EC_{50} of the chemical of interest. (B) K values at various concentrations of the synergist, calculated as the ratio of the EC_{50} of the targeted chemical in the absence of the synergist to the EC_{50} in the presence of a defined concentration of the synergist.

The response of the salmon to the toxicity of cypermethrin and the chemical mixture at the exposure site can be calculated from the data. First, the response to cypermethrin and deltamethrin at the exposure site is calculated with a concentration-addition model. Several concentration-addition model formats have been developed (Niederlehner et al. 1998; Safe 1998; Altenburger et al. 2000; Cleuvers 2003; Olmstead and LeBlanc 2005b) that are derivations of the original model presented by Finney (1942). Next, the responses (immobilization and loss of equilibrium) to each chemical, or groups of similarly acting chemicals, in the mixture are calculated with a response-addition model. The model originally described by Bliss (1939) remains the model of choice for calculating the joint action of dissimilarly acting chemicals (Backhaus et al. 2000; Walter et al. 2002; Olmstead and LeBlanc 2005b). Individual responses are then multiplied by appropriate K values to account for synergistic interactions. In this exercise, the K values of PBO are both 1.0; this indicates that the PBO concentration in the environment is not expected to modify the toxicity of cypermethrin or deltamethrin. Finally, the adjusted individual responses are summed to provide an estimate of the predicted response to the chemical mixture.

Data requirements and logistical considerations often temper efforts to assess the toxicity of chemical mixtures. Decisions on which chemicals in a mixture have the highest probability of toxic interaction with the pesticide active ingredient must be made and used to focus the assessment. Guidance for that decision-making is summarized in the section "Conclusions and Recommendations" of this chapter. Ultimately, some judgment must be made as to whether the possible adverse consequences associated with ignoring mixture effects outweigh the uncertainty associated with using nonempirical approaches to fill data gaps and to estimate the effects of mixture constituents on pesticide toxicity.

INTERSPECIES EXTRAPOLATIONS AND SURROGATE SPECIES

Different species often respond differently to chemical exposures because of differences in, for example, metabolic rates and pathways, the presence of functional genes, and different enzyme systems. Those differences can result in large differences in sensitivity; for example, adult guinea pigs (*Cavia porcellus*) are up to 5,000 times more sensitive to dioxins than are hamsters (*Mesocricetus auratus*) (Kociba and Schwetz 1982). Therefore, there is concern about how to extrapolate toxicity information from tested species to species of concern. Although the idea of finding a scientifically credible surrogate species might be appealing, the committee finds this approach difficult for two reasons.

First, it is not always straightforward to select a scientifically credible surrogate for a listed species. For example, rainbow trout are often used as surrogates for endangered Pacific salmon species because they are in the same genus, but they might respond differently to chemicals and other environmental stressors (Buhl and Hamilton 1991; EPA 2007). Moreover, there are seasonal differences in timing of breeding and larval development among species—generally,

Exposure

trout in the spring and salmon in the fall (Quinn 2005)—and differences among species in sensitivity during development, juveniles being more sensitive than larval fish (Buhl and Hamilton 1991). Consequently, exposure at a given place and time will have different effects among species because of their inherent vulnerability and differences in timing of development (see Box 4-1). There are many other physiological and ecological differences among species that will also affect their vulnerability to stressors, such as temperature and disease, so the overall effect of exposure to contaminants at a given time of the year can vary considerably.

Second, not all species are amenable to the degree of domestication required to conduct laboratory experiments, and the use of some might lead to public objections if they are suggested for such studies, for example, dogs as wolf surrogates or cats as jaguar or ocelot surrogates. It is also hard to imagine an appropriate surrogate for many species, such as polar or grizzly bears. Therefore, a scientifically credible alternative approach is to define a range of sensitivities within which the sensitivity of a species of concern could reasonably be expected to occur or a range that could be used to make reasoned extrapolations from species that have been tested by using inferences based on other chemicals. Life histories would need to be considered. If different life histories lead two related species that have similar toxicological sensitivities to a chemical to occupy different locations at different times, their susceptibility to a chemical could be quite different.

Listed species are not inherently more sensitive to chemicals than species that are not listed (Sappington et al. 2001; Besser et al. 2005; Dwyer et al. 2005), so similar methods of cross-species extrapolations can be used for any ecological risk assessment. Those methods include interspecies correlation analyses or interspecies correlation estimation (ICE) (Dyer 2006; Raimondo et al. 2010) and species sensitivity distributions (SSDs) (Posthuma et al. 2001). ICE models use the initial toxicity estimate for one species to estimate toxicity values for other species. The toxicity values can then be used either directly (if the species whose values are predicted is the species of concern) or in the development of SSDs. Knowledge about the mechanism of action of a chemical and the physiological similarity between test species and species of concern can provide empirical evidence to use in interpreting the theoretical relationships derived with the ICE model. Dyer et al. (2006) showed that using estimated values in an SSD in addition to or instead of measured values results in identification of threshold concentrations within an order of magnitude of those derived from distributions based only on empirical data. If a small dataset is available, bootstrapping or a Monte Carlo analysis may also be used to generate a response distribution (Warren-Hicks and Hart 2010). Bayesian approaches, which use all the information underlying the concentration-response data and result in presentation of the entire range of values that could be encountered beyond those of the tested species, have also been used to generate the end-point values for use in an SSD (Moore et al. 2010).

An SSD is a statistical distribution of the various concentrations at which different species have the same response to a chemical (Posthuma et al. 2001). Figure 4-4 provides an example of an SSD. The simplest approach is to display the SSD as a cumulative distribution function in much the same way that interindividual variability is displayed as an exposure-response function. An SSD can be based on any outcome—such as mortality, growth, or enzyme activity—for any group of species (such as all aquatic species, fishes, or plants) for any metric, such as EC_{50}, NOEC, or LC_{50}. Generally, a lognormal distribution is assumed, although Newman et al. (2000) point out that such an assumption is not always valid and that when sufficient data are available a data-specific distribution should be used. At the very least, the model's goodness of fit should be evaluated or acknowledged (Farrar et al. 2010). Furthermore, the data points used to generate the SSD have associated uncertainty that should be carried forward in generating the distribution. That uncertainty can be used to put confidence limits around the hazard concentration (HC) at the selected percentile and to determine the number of data points needed to define the HC with a desired amount of precision. Generally, the 5th percentile of the distribution is accepted as a matter of policy as the concentration that would maintain the viability of most species (HC_5), and preference is given to using the lower confidence limit of the HC_5. The HC_5s from SSDs of multiple chemicals for aquatic organisms—whether based on tested species or on extrapolations from ICE models—have been shown to be significantly lower than concentrations derived by using safety factors of 10 and 97% lower than the LC_{50}s of all endangered species (Raimondo et al. 2008).

A reasonable alternative to the use of SSD models is to use concentration-response models (or single-point estimates) available for each species to assign values to parameters in the population model with a Monte Carlo approach. For example, the percentage of the population that survives an estimated exposure can be randomly selected from all the species tested. At the exposure concentration of interest, survival might be 50% for a population of quail, 75% for a songbird, and 30% for a duck. One of those values would be randomly selected, and the population model would be run to determine the population-level end point (for example, lambda or risk of extinction or decline); that process would be repeated 1,000 times to generate a distribution of the population-level end points that reflects the range of possible survival rates of the nontested species at the estimated exposure concentrations. That approach assumes that the species of interest has an unknown survival rate that is encompassed by the range of survival rates of all other measured species. The latter species could be constrained to ones that are similar taxonomically (or physiologically) to the species of concern, or all species could be included to make the resulting analysis as robust as possible. All types of population-level end-point values (such as median, mean, and upper or lower bounds) are carried forward to the risk characterization. The process outlined here would be conducted simultaneously for reproduction.

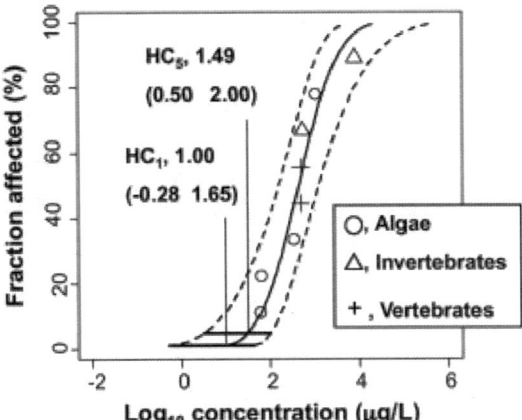

FIGURE 4-4 Species sensitivity distributions for 2,2'-dipyridyldisulfide derived by using a Bayesian statistical model. Source: Mochida et al. 2012. Reprinted with permission; copyright 2012, *Aquatic Toxicology*.

The committee concludes that the procedures outlined above, which result in a range of sensitivities, are good alternatives to the use of a single surrogate species. The use of a single surrogate species is often difficult to justify, but the use of a set of species would give a good idea of the range of possible organismal responses to a pesticide. As noted, life-history variations would need to be considered.

OTHER UNCERTAINTIES IN EFFECTS ANALYSIS

An effects analysis describes how a species of concern or a biological component of its habitat responds when exposed to a stressor, in this case a pesticide formulation, including the active ingredient and other constituent chemicals. It also includes an analysis of how the pesticide interacts with other environmental stressors, either increasing or decreasing the toxic response. However, all such estimates are uncertain, because of either measurement error or lack of knowledge. As stated in Chapter 2, the uncertainty should be clearly defined and propagated through the risk assessment. Currently, EPA and the Services do not quantitatively incorporate uncertainty in the effects analysis. Although they all report confidence intervals around most of the effects end points, they ultimately use only deterministic approaches (single point estimates of the magnitude of effect at a particular exposure concentration) or qualitative descriptions (particularly for behavioral and sublethal effects other than quantifiable reproduction responses). And they do not explain their selection of particular effects concentrations when selecting toxicity thresholds (for example, the choice of an EC_{25} instead of an EC_{10}, or vice versa). Therefore, much quantitative information that could explain the possible range of effects at single or multiple exposure concentrations is not used. General statements about "uncertainty" or "considerable uncertainty" are made, and these provide little or no guidance to a decision-maker because the terms are vague and open to multiple interpretations.

The committee recommends that all parties use quantitative uncertainty analyses whenever sufficient data are available. Selection of a particular approach depends on the amount of data available, the timeframe for conducting an assessment, and the preference of the risk assessors. Nearly all toxicity studies provide some information about concentration-response relationships, including measures of variability; and many measured environmental-response variables, such as productivity rates, also have time-dependent variability estimates. Therefore, quantified uncertainty estimates about effects values can be developed and used in the risk assessments. There are many well-documented methods for quantifying uncertainty in chemical toxicity assessments and in population models that are supported by user-friendly commercial software, including probability bounds, confidence intervals, Monte Carlo analysis, and Bayesian techniques (Spear 1997; Borsuk et al. 2004; Solomon et al. 2008; Link and Barker 2010; Warren-Hicks and Hart 2010; McLaughlin and Jain 2011). If quantitative effects data are insufficient for input into a quantified risk assessment or a quantitative uncertainty analysis, a qualitative statement about the risk potential and degree of uncertainty (such as low, medium, or high) can be used instead provided that they are accompanied by some bounding definitions (such as different by orders of magnitude). In cases in which some effects have quantifiable uncertainty and others do not, the committee recommends that the formal risk assessment focus on end points that are quantifiable and include quantitative measures of variability. The "others" should be described qualitatively and used as supplemental information after qualitative uncertainty categories or lines of evidence that might be useful are clearly defined.

CONCLUSIONS AND RECOMMENDATIONS

Sublethal, Indirect, and Cumulative Effects

- An adverse effect should be defined by the degree to which an organism's survival or reproduction is affected; thus, assessing the effects of a pesticide on a listed species requires quantifying the effect of the pesticide on survival and reproduction of the species in the wild. Any effect that results in a change in survival or reproduction is relevant to the assessment, and any effect that does not change either outcome is irrelevant with respect to a quantitative assessment of population effects.
- To determine whether a pesticide is "likely to adversely affect" a listed species, a broad search should be conducted to identify information on sublethal effects of the pesticide and possible concentration-response relationships.

Exposure

- To provide information to support a jeopardy determination, the Services should either (a) show how sublethal effects change survival or reproduction and incorporate such information into the population viability analysis or (b) state that such relationships are unknown but possible and include a qualitative discussion of uncertainty in the BiOp.
- When indirect effects can be quantified, they should be incorporated into the effects analysis.
- Population models provide an appropriate framework for incorporating baseline conditions and projected future cumulative effects into the assessment. Evaluation of those effects is within the purview of the Services and an appropriate part of the BiOp.

Effects Models

- Because the ESA is concerned with species or listed units within named species, the effects of pesticides must be expressed at the population level. Accordingly, population models that incorporate temporal variability and focus on probabilistic results are needed for assessing population risks. Although deterministic projection models are insufficient for that task, they can be used in the absence of information on temporal variability in the elements of the baseline model provided that the risk assessment recognizes the potential bias that might result from using them.
- Spatial structure and density dependence might have important effects on population dynamics and must be incorporated into population models when data are available. However, in the absence of such data, it is appropriate to use generic, single-population models that characterize the life history of a group of species to estimate the effects of a pesticide on a given species.
- The assumption that mortality due to pesticide exposure will always be compensated for by density dependence is not scientifically valid because such exposure will likely decrease the growth rate of the population at all densities and generally depress the population growth-density curve.
- For the purposes of population modeling, effects need to be estimated at a range of concentrations that includes all values that the population might reasonably encounter. Test results expressed only as threshold values or point estimates—such as the no-observed-adverse-effects level, the lowest observed-adverse-effects level, and the LC_{50}—do not provide enough information for a population-level risk assessment.

Mixtures

- A quantitative mixture risk assessment requires extensive data, including data on the identity, concentration, and toxicity of mixture components. Challenges in assessing risk to listed species from pesticide-containing mixtures arise largely because of the lack of such data *and* the lack of understanding of the potential for interactions among mixture components.

- In the absence of such quantitative data, the possible contribution of specific mixture components to the toxicity of a pesticide active ingredient cannot be incorporated into a quantitative risk assessment. However, the risk assessor should describe the possible effects of mixture components on the risk estimate to the decision-maker.
- The committee emphasizes that the complexity of assessing the risk posed by chemical mixtures should not paralyze the process. The following guidelines provide a tool for helping to determine when and how to consider components other than the pesticide active ingredient in a risk assessment:

1. The toxicity of the pesticide active ingredient is central to the assessment. Other chemicals are relevant only if they modify the toxicity of the pesticide active ingredient or the susceptibility of the species of concern to the active ingredient.
2. The toxicity end point most relevant to the species of concern must be determined before initiation of the effects analysis.
3. Mixture components that do not elicit the relevant response in the species of concern do not need to be considered in the effects analysis. Mixture components that do elicit the relevant response need to be considered in the effects analysis.
4. In the absence of any data that would support the hypothesis of a synergistic interaction between the pesticide active ingredient and other mixture components, the effects analysis should proceed on the assumption that the components have additive effects.
5. For chemicals that have common mechanisms of action and parallel slopes in the concentration-response curves, concentration addition is a reasonable approach for modeling additive effects. However, caution should be exercised in using concentration-addition modeling as a default approach when no mechanistic data or concentration-response data are available.
6. For chemicals that have different mechanisms of action, response addition (with a zero correlation of individual tolerances) is a reasonable approach for modeling additive effects. For this case, mixture components will contribute to the response only when present in the environment at concentrations that elicit the relevant response. That is, such components do not need to be considered when present at concentrations below their toxic thresholds.
7. Potential synergistic interactions need be considered only when a synergist is present in the environment above its interaction threshold concentration. In the case of synergism, it is probably prudent to generate information on toxic interactions to ensure accurate evaluation of potential responses of the species of concern.

8. In the case of antagonism, uncertainties associated with both exposures and toxic interactions will seldom justify a quantitative modification of the effects analysis.
9. The use of uncertainty factors to compensate for the absence of information on potential interactions of mixture components is not recommended. When data are available, quantitative methods can be used to evaluate the interactions.

Interspecies Extrapolation, Surrogate Species, and Other Uncertainties

- Many difficulties are associated with the use of surrogate species to estimate risk to a species on which data are not available or cannot easily be obtained.
- An alternative approach to using a single surrogate species is to define a range of sensitivities within which the sensitivity of the species of concern could reasonably be expected to occur or a range of sensitivities that could be used to make reasoned extrapolations from species that have been tested by using inferences based on other chemicals.
- Because listed species are not inherently more sensitive to chemicals than species that are not listed, similar methods of cross-species extrapolations can be used for any ecological risk assessment and include interspecies correlation analysis and species sensitivity distributions.
- Life histories need to be considered whether one is identifying a single surrogate species or using an alternative approach. For example, if two related species have similar toxic sensitivities to a chemical but have different life histories that lead them to occupy different locations at different times, their susceptibility to the chemical could be different.
- In all cases and for all methods, quantitative uncertainty analyses should be used whenever sufficient data are available.

REFERENCES

Adam, O., P.M. Badot, F. Degiorgi, and G. Crini. 2009. Mixture toxicity assessment of wood preservative pesticides in the fresh water amphipod *Gammarus pulex* (L.). Ecotoxicol. Environ. Saf. 72(4):441-449.

Akçakaya, H.R., M.A. Burgman, and L. Ginzburg. 1999. Applied Population Ecology Principles and Computer Exercises using RAMAS EcoLab 2.0, 2nd Ed. Sunderland, MA: Sinauer Associates.

Akçakaya, H.R., J.D. Stark, and T.S. Bridges, eds. 2008. Demographic Toxicity: Methods in Ecological Risk Assessment. New York: Oxford University Press.

Altenburger, R., T. Backhaus, W. Boedeker, M. Faust, M. Scholze, and L.H. Grimme. 2000. Predictability of the toxicity of multiple chemical mixtures to *Vibrio fischeri*: Mixtures composed of similarly acting chemicals. Environ. Toxicol. Chem. 19(9):2341-2347.

Amweg, E.L., D.P. Weston, J. You, and M.J. Lydy. 2006. Pyrethroid insecticides and sediment toxicity in urban creeks from California and Tennessee. Environ. Sci. Technol. 40(5):1700-1706.

Anderson, T.D., and K.Y. Zhu. 2004. Synergistic and antagonistic effects of atrazine on the toxicity of organophosphorodithioate and organophosphorothioate insecticides to *Chironomus tentans* (Diptera: *Chironomidae*). Pestic. Biochem. Physiol. 80(1):64-64.

Ankley, G.T., and S.A. Collyard. 1995. Influence of piperonyl butoxide on the toxicity of organophosphate insecticides to three species of freshwater benthic invertebrates. Comp. Biochem. Physiol. C Pharmacol. Toxicol. Endocrinol. 110(2):149-155.

Backhaus, T., R. Altenburger, W. Boedeker, M. Faust, M. Scholze, and L.H. Grimme. 2000. Predictability of the toxicity of a multiple mixture of dissimilarly acting chemicals to *Vibrio fischeri*. Environ. Toxicol. Chem. 19(9):2348-2356.

Backhaus, T., A. Arrhenius, and H. Blanck. 2004. Toxicity of a mixture of dissimilarly acting substances to natural algal communities: Predictive power and limitations of independent action and concentration addition. Environ. Sci. Technol. 38(23):6363-6370.

Baird, D.J., L. Maltby, P.W. Greig-Smith, and P.E.T. Douben. 1996. ECOtoxicology: Ecological Dimensions. London: Chapman & Hall.

Baldwin, D.H., J.A. Spromberg, T.K. Collier, and N.L. Scholz. 2009. A fish of many scales: Extrapolating sublethal pesticide exposures to the productivity of wild salmon populations. Ecol. Appl. 19(8):2004-2015.

Barnthouse, L.W., W.R. Munns, and M.T. Sorensen. 2007. Population-Level Ecological Risk Assessment. Pensacola, FL: CRC Press.

Barry, M.J. 1998. Endosulfan-enhanced crest induction in *Daphnia longicephala*: Evidence for cholinergic innervation of kairomone receptors. J. Plankton Res. 20(7):1219-1231.

Beckman, B.R., D.A. Larsen, B. Lee-Pawlak, and W.W. Dickhoff. 1998. Relation of fish size and growth rate to migration of spring chinook salmon smolts. N. Am. J. Fish. Manage. 18(3):537-546.

Behnke, R.J. 1992. Native Trout of Western North America. American Fisheries Society Monograph 6. December 1992.

Belden, J.B., R.J. Gilliom, J.B. Martin, and M.J. Lydy. 2007a. Relative toxicity and occurrence patterns of pesticide mixtures in streams draining agricultural watersheds dominated by corn and soybean production. Integr. Environ. Assess. Manag. 3(1):90-100.

Belden, J.B., R.J. Gilliom, and M.J. Lydy. 2007b. How well can we predict the toxicity of pesticide mixtures to aquatic life? Integr. Environ. Assess. Manag. 3(3):364-372.

Besser, J.M., N. Wang, F.J. Dwyer, F.L. Mayer, Jr., and C.G. Ingersoll. 2005. Assessing contaminant sensitivity of endangered and threatened aquatic species, Part II. Chronic toxicity of copper and pentachlorophenol to two endangered species and two surrogate species. Arch. Environ. Contam. Toxicol. 48(2):155-165.

Beyond Pesticides. 2009. Re: Registration Review; Glyphosate Docket Opened for Review and Comment. Docket Number: EPA-HQ-OPP-2009-0361 [online]. Available: http://www.beyondpesticides.org/documents/glyphosate-final9-21-1.pdf [accessed Nov. 16, 2012].

Bielza, P., P.J. Espinosa, V. Quinto, J. Abellán, and J. Contreras. 2007. Synergism studies with binary mixtures of pyrethroid, carbamate and organophosphate insecticides on *Frankliniella occidentalis* (Pergande). Pest. Manag. Sci. 63(1):84-89.

Bjergager, M.B., M.L. Hanson, K.R. Solomon, and N. Cedergreen. 2012. Synergy between prochloraz and esfenvalerate in *Daphnia magna* from acute and subchronic exposures in the laboratory and microcosms. Aquat. Toxicol. 110-111:17-24.

Bliss, C.I. 1939. The toxicity of poisons applied jointly. Ann. Appl. Biol. 26(3):585-615.

Boobis, A., R. Budinsky, S. Collie, K. Crofton, M. Embry, S. Felter, R. Hertzberg, D. Kopp, G. Mihlan, M. Mumtaz, P. Price, K. Solomon, L. Teuschler, R. Yang, and R. Zaleski. 2011. Critical analysis of literature on low-dose synergy for use in screening chemical mixtures for risk assessment. Crit. Rev. Toxicol. 41(5):369-383.

Borsuk, M.E., C.H. Stow, and K.H. Reckhow. 2004. A Bayesian network of eutrophication models for synthesis, prediction, and uncertainty analysis. Ecol. Model. 173(2-3):219-239.

Buhl, K.J., and S.J. Hamilton. 1991. Relative sensitivity of early life stages of arctic grayling, coho salmon, and rainbow trout to nine inorganics. Ecotox. Environ. Safe 22(2):184-197.

Burgman, M.A., S. Ferson, and H.R. Akçakaya. 1993. Risk Assessment in Conservation Biology. London: Chapman and Hall.

Cairns Jr., J., and J. Pratt. 1993. Trends in ecotoxicology. Sci. Total Environ. 134(suppl. 1):7-22.

Cashman, J.R., B.Y. Perotti, C.E. Berkman, and J. Lin. 1996. Pharmacokinetics and molecular detoxication. Environ. Health Perspect. 104(Suppl. 1):23-40.

Caswell, H. 2001. Matrix Population Models: Construction, Analysis, and Interpretation, 2nd Ed. Sunderland, MA: Sinauer Associates.

Cedergreen, N., A. Kamper, and J.C. Streibig. 2006. Is prochloraz a potent synergist across aquatic species? A study on bacteria, daphnia, algae and higher plants. Aquat. Toxicol. 78(3):243-252.

Clement, J.G. 1984. Role of aliesterase in organophosphate poisoning. Fundam. Appl. Toxicol. 4(2 Pt. 2):S96-S105.

Cleuvers, M. 2003. Aquatic ecotoxicity of pharmaceuticals including the assessment of combination effects. Toxicol. Lett. 142(3):185-194.

Coors, A., and L. De Meester. 2008. Synergistic, antagonistic and additive effects of multiple stressors: Predation threat, parasitism, and pesticide exposure in *Daphnia magna*. J. Appl. Ecol. 45(6):1820-1828.

Courchamp, F., J. Berec, and J. Gascoigne. 2008. Allee Effects in Ecology and Conservation. New York: Oxford University Press.

Dallegrave, E., F.D. Mantese, R.T. Oliveira, A.J. Andrade, P.R. Dalsenter, and A. Langeloh. 2007. Pre- and postnatal toxicity of the commercial glyphosate formulation in Wistar rats. Arch. Toxicol. 81(9):665-673.

Dobrev, I.D., M.E. Andersen, and R.S. Yang. 2001. Assessing interaction thresholds for trichloroethylene in combination with tetrachloroethylene and 1,1,1-trichloroethane using gas uptake studies and PBPK modeling. Arch. Toxicol. 75(3):134-144.

Duchet, C., M.A. Coutellec, E. Franquet, C. Lagneau, and L. Lagadic. 2010. Population-level effects of spinosad and *Bacillus thuringiensis israelensis* in *Daphnia pulex* and *Daphnia magna*: Comparison of laboratory and field microcosm exposure conditions. Ecotoxicology 19(7):1224-1237.

Duffy, E.J., and D.A. Beauchamp. 2008. Seasonal patterns of predation on juvenile Pacific salmon by anadromous cutthroat trout in Puget Sound. T. Am. Fish. Soc. 137(1):165-181.

Dwyer, F.J., F.L. Mayer, L.C. Sappington, D.R. Buckler, C.M. Bridges, I.E. Greer, D.K. Hardesty, C.E. Henke, C.G. Ingersoll, J.L. Kunz, D.W. Whites, T. Augspurger, D.R. Mount, K. Hattala, and G.N. Neuderfer. 2005. Assessing contaminant sensitivity of endangered and threatened aquatic species: Part I. Acute toxicity of five chemicals. Arch. Environ. Contam. Toxicol. 48(2):143-154.

Dyer, S.D., D.J. Versteeg, S.E. Belanger, J.G. Chaney, and F.L. Mayer. 2006. Interspecies correlation estimates predict protective environmental concentrations. Environ. Sci. Technol. 40(9):3102-3131.

El-Masri, H. 2010. Toxicological interaction thresholds of chemical mixtures. Pp. 207-214 in Principles and Practice of Mixtures Toxicology, M. Mumtaz, ed. Weinheim: Wiley-VCH.

El-Masri, H.A., K.F. Reardon, and R.S. Yang. 1997. Integrated approaches for the analysis of toxicologic interactions of chemicals mixtures. Crit. Rev. Toxicol. 27(2):175-197.

El-Merhibi, A., A. Kumar, and T. Smeaton. 2004. Role of piperonyl butoxide in the toxicity of chlorpyrifos to *Ceriodaphnia dubia* and *Xenopus laevis*. Ecotoxicol. Environ. Saf. 57(2):202-212.

EPA (U.S. Environmental Protection Agency). 1986. Guidelines for the Health Risk Assessment of Chemical Mixtures. EPA/630/R-98/002. Risk Assessment Forum, U.S. Environmental Protection Agency, Washington, DC. September 1986 [online]. Available: http://www.epa.gov/raf/publications/pdfs/CHEMMIX_1986.pdf [accessed Nov. 19, 2012].

EPA (U.S. Environmental Protection Agency). 1989. Interim Procedures for Estimating Risks Associated with Exposures to Mixtures of Chlorinated Dibenzo-p-Dioxins and Dibenzofurans (CDDs and CDFs) and 1989 Update. EPA/625/3-89/016. NTIS PB90145756. Risk Assessment Forum, U.S. Environmental Protection Agency, Washington, DC.

EPA (U.S. Environmental Protection Agency). 1996. Health Effects Test Guidelines OPPTS 870.1200 Acute Dermal Toxicity. EPA 712-C-96-192. Office for Prevention, Pesticides and Toxic Substances, U.S. Environmental Protection Agency. June 1996 [online]. Available: http://iccvam.niehs.nih.gov/SuppDocs/FedDocs/EPA/EPA-870-1200.pdf [accessed Nov. 20, 2012].

EPA (U.S. Environmental Protection Agency). 1998a. Guidelines for Ecological Risk Assessment. EPA/630/R-95/002F. Risk Assessment Forum, U.S. Environmental Protection Agency, Washington, DC. April 1998 [online]. Available: http://www.epa.gov/raf/publications/pdfs/ECOTXTBX.PDF [accessed Nov. 19, 2012].

EPA (U.S. Environmental Protection Agency). 1998b. Health Effects Test Guidelines OPPTS 870.1100 Acute Oral Toxicity. EPA 712-C-98-190. Office for Prevention, Pesticides and Toxic Substances, U.S. Environmental Protection Agency. August 1998 [online]. Available: http://iccvam.niehs.nih.gov/methods/acutetox/invidocs/EPA_870_1100.pdf [accessed Nov. 2012].

EPA (U.S. Environmental Protection Agency). 1998c. Health Effects Test Guidelines OPPTS 870.1300 Acute Inhalation Toxicity. EPA 712-C-98-193. Office for Prevention, Pesticides and Toxic Substances, U.S. Environmental Protection Agency. August 1998 [online]. Available: http://iccvam.niehs.nih.gov/SuppDocs/FedDocs/EPA/EPA-870-1300.pdf [accessed Nov. 20, 2012].

EPA (U.S. Environmental Protection Agency). 1998d. Health Effects Test Guidelines OPPTS 870.2400 Acute Eye Irritation. EPA 712-C-98-195. Office for Prevention, Pesticides and Toxic Substances, U.S. Environmental Protection Agency. August 1998 [online]. Available: http://iccvam.niehs.nih.gov/SuppDocs/FedDocs/EPA/EPA_870_2400.pdf [accessed Nov. 20, 2012].

EPA (U.S. Environmental Protection Agency). 1998e. Health Effects Test Guidelines OPPTS 870.2500 Acute Dermal Irritation. EPA 712-C-98-196. Office for Prevention, Pesticides and Toxic Substances, U.S. Environmental Protection Agency.

Exposure 139

August 1998 [online]. Available: http://iccvam.niehs.nih.gov/SuppDocs/FedDocs/ EPA/EPA_870_2500.pdf [accessed Nov. 20, 2012].

EPA (U.S. Environmental Protection Agency). 2000. Supplementary Guidance for Conducting Health Risk Assessment of Chemical Mixtures. EPA/630/R-00/002. Risk Assessment Forum Technical Panel, U.S. Environmental Protection Agency, Washington, DC. August 2000 [online]. Available: http://www.epa.gov/raf/publications/pdfs/CHEM_MIX_08_2001.PDF [accessed Nov. 19, 2012].

EPA (U.S. Environmental Protection Agency). 2002. Guidance on Cumulative Risk Assessment of Pesticide Chemicals That Have a Common Mechanism of Toxicity. Office of Pesticide Programs, U.S. Environmental Protection Agency, Washington, DC. January 14, 2002 [online]. Available: http://www.epa.gov/oppfead1/trac/science/cumulative_guidance.pdf [accessed Nov. 19, 2012].

EPA (U.S. Environmental Protection Agency). 2003. Health Effects Test Guidelines OPPTS 870.2600 Skin Sensitization. EPA 712-C-03-197. Office for Prevention, Pesticides and Toxic Substances, U.S. Environmental Protection Agency. March 2003 [online]. Available: http://www.regulations.gov/#!documentDetail;D=EPA-HQ-OPPT-2009-0156-0008 [accessed Nov. 20, 2012].

EPA (U.S. Environmental Protection Agency). 2004. Overview of the Ecological Risk Assessment Process in the Office of Pesticide Programs, U.S. Environmental Protection Agency: Endangered and Threatened Species Effects Determinations. Office of Prevention, Pesticides and Toxic Substances, Office of Pesticide Programs, U.S. Environmental Protection Agency, Washington DC. January 23, 2004 [online]. Available: http://www.epa.gov/oppfead1/endanger/consultation/ecorisk-overview.pdf [accessed Aug. 29, 2012].

EPA (U.S. Environmental Protection Agency). 2005. Overview of the Piperonyl Butoxide Risk Assessments. Docket ID EPA-HQ-OPP-2005-0042. Pesticides: Reregistration, U.S. Environmental Protection Agency [online]. Available: http://www.epa.gov/opp00001/reregistration/piperonyl/ [accessed Feb. 19, 2013].

EPA (U.S. Environmental Protection Agency). 2007. Aquatic Life Ambient Freshwater Quality Criteria – Copper, 2007 Revision. EPA-822-R-07-001. Office of Water, Office of Science and Technology, U.S. Environmental Protection Agency, Washington, DC. February 2007 [online]. Available: http://water.epa.gov/scitech/swguidance/standards/criteria/aqlife/copper/upload/2009_04_27_criteria_copper_2007_criteria-full.pdf [accessed Feb. 19, 2013].

EPA (U.S. Environmental Protection Agency). 2008. Risks of Glyphosate Use to Federally Threatened California Red-legged Frog (*Rana aurora draytonii*). Pesticide Effects Determination. Environmental Fate and Effects Division, Office of Pesticide Programs, Washington, DC. October 17, 2008 [online]. Available: http://www.epa.gov/espp/litstatus/effects/redleg-frog/glyphosate/determination.pdf [accessed on Feb. 26, 2012].

EPA (U.S. Environmental Protection Agency). 2012a. EPA Response to NAS Questions, March 28, 2012 [online]. Available: http://www.thecre.com/forum1/wp-content/uploads/2012/04/ESAEPAATRAZINE.pdf [accessed Nov. 15, 2012].

EPA (U.S. Environmental Protection Agency). 2012b. Inert Ingredients in Pesticide Products-Reassessment Decision Documents. U.S. Environmental Protection Agency [online]. Available: http://www.epa.gov/opprd001/inerts/decisiondoc_a2k.html [accessed Mar. 26, 2012].

EPA (U.S. Environmental Protection Agency). 2012c. General Guidance for Petitioning the Agency for the Establishment of a New/Amended Food Use Inert Ingredient Tolerance or Tolerance Exemption. Office of Chemical Safety and Pollution Prevention, Office of Pesticide Programs, Registration Division, U.S. Environmental Protection Agency. September 12, 2012 [online]. Available: www.epa.gov/opprd001/inerts/inertpetition.pdf [accessed Nov. 19, 2012].

EPA (U.S. Environmental Protection Agency). 2012d. General Guidance for Requesting a New Nonfood Use Inert Ingredient. Office of Chemical Safety and Pollution Prevention, Office of Pesticide Programs, Registration Division, U.S. Environmental Protection Agency. September 13, 2012. [online]. Available: http://www.epa.gov/opprd001/inerts/nonfood_inert.pdf [accessed Nov. 19, 2012].

EPA (U.S. Environmental Protection Agency). 2012e. Guidance to Petitioners for Low Risk Polymer Submissions. Office of Chemical Safety and Pollution Prevention, Office of Pesticide Programs, Registration Division, U.S. Environmental Protection Agency. September 12, 2012 [online]. Available: http://www.epa.gov/opprd001/inerts/lowriskpolymer.pdf [accessed Nov. 19, 2012].

EPA (U.S. Environmental Protection Agency). 2012f. Pesticide Registration Manual, Chapter 2: Registering a Pesticide Product. May 2012 [online]. Available: http://www.epa.gov/opprd001/registrationmanual/chapter2.html#metoo [accessed Nov. 19, 2012].

EPA (U.S. Environmental Protection Agency). 2012g. Guidance for Waiving or Bridging of Mammalian Acute Toxicity Tests for Pesticides and Pesticide Products (Acute Oral, Acute Dermal, Acute Inhalation, Primary Eye, Primary Dermal, and Dermal Sensitization). Office of Pesticide Programs. Available: http://www.epa.gov/opp00001/science/acute-data-waiver-guidance.pdf [accessed Nov. 19, 2012].

EPA (U.S. Environmental Protection Agency). 2013. Harmonized Test Guidelines. U.S. Environmental Protection Agency [online]. Available: http://www.epa.gov/ocspp/pubs/frs/home/guidelin.htm [accessed Mar. 26, 2013].

Farrar, D., T. Barry, P. Hendley, M. Crane, P. Mineau, M.H. Russell, and E.W. Odenkirchen. 2010. Issues underlying the selection of distributions. Pp. 31-52 in Application of Uncertainty Analysis to Ecological Risks of Pesticides, W.J. Warren-Hicks, and A. Hart, eds. Pensacola, FL: SETAC Press.

Feron, V.J., J.P. Groten, D. Jonker, F.R. Cassee, and P.J. van Bladeren. 1995. Toxicology of chemical mixtures: Challenges for today and the future. Toxicology 105(2-3): 415-427.

Ferson, S., L.R. Ginzburg, and R.A. Goldstein. 1996. Inferring ecological risk from toxicity bioassays. Water Air Soil Poll. 90(1-2):71-82.

FESTF (FIFRA Endangered Species Task Force). 2012. FIFRA Endangered Species Task Force Data Use in the Assessment of the Risk of FIFRA-Regulated Pesticides to Species Regulated by the ESA. A background document prepared by the FIFRA Endangered Species Task Force for the NAS Panel on Ecological Risk Assessment under FIFRA and ESA, January 25, 2012.

Finney, D.J. 1942. The analysis of toxicity tests on mixtures of poisons. Ann. Appl. Biol. 29(1):82-94.

Exposure 141

Finney, D.J. 1971. Probit Analysis, 3rd Ed. Cambridge: Cambridge University Press.

Forbes, V.E., and P. Calow. 1999. Is the per capita rate of increase a good measure of population-level effects in ecotoxicology? Environ. Toxicol. Chem. 18(7):1544-1556.

Forbes, V.E., R.M. Sibly, and P. Calow. 2001. Toxicant impacts on density-limited populations: A critical review of theory, practice, and results. Ecol. Appl. 11(4):1249-1257.

Forbes, V.E., R.M. Sibly, and I. Linke-Gamenick. 2003. Joint effects of population density and toxicant exposure on population dynamics of *Capitella* sp. I. Ecol. Appl. 13(4):1094-1103.

FWS (U.S. Fish and Wildlife Service). 2009. Informal Consultation on the Effects of Diazinon and Racemic Metolachlor Reregistration on the Endangered Barton Springs Salamander. Letter to Arthur-Jean B. Williams, Environmental Fate and Effects Division, Office of Pesticide Programs, U.S. Environmental Protection Agency, from M.A. Nelson, Division of Consultation, Habitat Conservation Planning, Recovery and State Grants, U.S. Fish and Wildlife Service, Washington, DC. January 14, 2009.

FWS (U.S. Fish and Wildlife). 2012. Response to Committee Questions for Agency and Services, from Rick Sayers, Chief, Division of Consultation, HCPs, Recovery, and State Grants, U.S. Fish and Wildlife Service Endangered Species Program. March 28, 2012 [online]. Available: http://www.thecre.com/forum1/wp-content/uploads/2012/04/esafwsatrazine1.pdf [accessed Nov. 19, 2012].

Giesy, J.P., S. Dobson, and K.R. Solomon. 2000. Ecotoxicological risk assessment for Roundup herbicide. Rev. Environ. Contamin. Toxicol. 167:35-120.

Giorgi, A.E., T.W. Hillman, J.R. Stevenson, S.G. Hays, and C.M. Peven. 1997. Factors that influence the downstream migration rates of juvenile salmon and steelhead through the hydroelectric system in the mid-Columbia River basin. N. Am. J. Fish. Manage. 17(2):268-282.

Gustafson, R.G., R.S. Waples, J.M. Myers, L.A. Weitkamp, G.J. Bryant, O.W. Johnson, and J.J. Hard. 2007. Pacific salmon extinctions: Quantifying lost and remaining diversity. Conserv. Biol. 21(4):1009-1020.

Healey, M.C. 1991. Life history of chinook salmon (*Oncorhynchus tshawytscha*). Pp. 311-393 in Pacific Salmon Life Histories, C. Groot, and L. Margolis, eds. Vancouver: University of British Columbia Press.

Hilborn, R., T.P. Quinn, D.E. Schindler, and D.E. Rogers. 2003. Biocomplexity and fisheries sustainability. Proc. Natl. Acad. Sci. USA 100(11):6564-6568.

Hodgson, E., and P.E. Levi. 2001. Metabolism of pesticides. Pp. 531-562 in Handbook of Pesticide Toxicology, 2nd Ed., R.I. Krieger, ed. London: Academic Press.

IPCS (International Programme on Chemical Safety). 2009. Assessment of Combined Exposures to Multiple Chemicals: Report of a WHO/IPCS International Workshop on Aggregate/Cumulative Risk Assessment. Harmonization Project Document No. 7. Geneva: World Health Organization [online]. Available: http://www.who.int/ipcs/methods/harmonization/areas/workshopreportdocument7.pdf [accessed Mar. 26, 2013].

Jones, D.G. 1998. Piperonyl Butoxide: The Insecticide Synergist. New York: Academic Press.

Kociba, R.J., and B.A. Schwetz. 1982. Toxicity of 2,3,7,8-Tetrachlorodibenzo-p-dioxin (TCDD). Drug Metab. Rev. 13(3): 387-406.

Kolpin, D.W., D.T. Furlong, M.T. Meyer, E.M. Thurman, S.D. Zaugg, L.B. Barber, and H.T. Buxton. 2002. Pharmaceuticals, hormones, and other organic wastewater contaminants in U.S. streams, 1999-2000: A national reconnaissance. Environ. Sci. Technol. 36(6):1202-1211.

Kortenkamp, A., R. Evans, M. Faust, F. Kalberlah, M. Scholze, and U. Schuhmacher-Wolz. 2012. Investigation of the State of the Science on Combined Actions of Chemicals in Food through Dissimilar Modes of Action and Proposal for Science-based Approach for Performing Related Cumulative Risk Assessment. European Food Safety Authority, January 31, 2012 [online]. Available: http://www.efsa.europa.eu/en/supporting/doc/232e.pdf [accessed Mar. 26, 2013].

Kuhn, A., W.R. Munns, S. Poucher, D. Champlin, and S. Lussier. 2000. Prediction of population-level response from mysid toxicity test data using population modeling techniques. Environ. Toxicol. Chem. 19(9):2364-2371.

Labenia, J.S., D.H. Baldwin, B. L. French, J.W. Davis, and N.L. Scholz. 2007. Behavioral impairment and increased predation mortality in cutthroat trout exposed to carbaryl. Mar. Ecol. Prog. Ser. 329:1-11.

Laetz, C.A., D.H. Baldwin, T.K. Collier, V. Hebert, J.D. Stark, and N.L. Scholz. 2009. The synergistic toxicity of pesticide mixtures: Implications for risk assessment and the conservation of endangered Pacific salmon. Environ. Health Perspect. 117(3):348-353.

LeBlanc, L.A., J.L. Orlando, and K.M. Kuivila. 2004. Pesticide Concentrations in Water and in Suspended and Bottom Sediments in the New and Alamo Rivers, Salton Sea Watershed, California, April 2003. Data Series 104. U.S. Geological Survey, Reston, VA [online]. Available: http://swrcb2.swrcb.ca.gov/water_issues/programs/?swamp/docs/reglrpts/rb7_pesticide_saltonseaws.pdf [accessed Nov. 19, 2012].

Levine, T.E. 1996. The regulation of inert ingredients in the United States. Pp. 1-11 in Pesticide Formulation and Adjuvant Technology, C.L. Foy, and D.W. Pritchard, eds. Boca Raton, FL: CRC Press.

Link, W.A., and R.J. Barker. 2010. Bayesian Inference: With Ecological Applications. New York: Academic Press.

Macneale, K.H., P.M. Kiffney, and N.L Scholz. 2010. Pesticides, aquatic food webs, and the conservation of Pacific salmon. Front. Ecol. Environ. 8(9):475-482.

Mangel, M., and W.H. Satterthwaite. 2008. Combining proximate and ultimate approaches to understand life history variation in salmonids with application to fisheries, conservation, and aquaculture. Bull. Mar. Sci. 83(1):107-130.

Martin, T., O.G. Ochou, M. Vaissayre, and D. Fournier. 2003. Organophosphorus insecticides synergize pyrethroids in the resistant strain of cotton bollworm, *Helicoverpa armigera* (Hubner) (Lepidoptera: Noctuidae) from West Africa. J. Econ. Entomol. 96(2):468-474.

McLaughlin, D.B., and V. Jain. 2011. Using Monte Carlo analysis to characterize the uncertainty in final acute values derived from aquatic toxicity data. Int. Environ. Assess. Manag. 7(2):269-279.

Mebane, C.A., and D.L. Arthaud. 2010. Extrapolating growth reductions in fish to changes in population extinction risks: Copper and Chinook salmon. Hum. Ecol. Risk Assess. 16(5):1026-1065.

Mochida, K., H. Amano, K. Ito, M. Ito, T. Onduka, H. Ichihashi, A. Kakuno, H. Harino, and K. Fujii. 2012. Species sensitivity distribution approach to primary risk analysis of the metal pyrithione photodegradation product, 2,2′-dipyridyldisulfide in the Inland Sea and induction of notochord undulation in fish embryos. Aqua. Toxicol. 118-119:152-163.

Moe, S.J. 2007. Density dependence in ecological risk assessment. Pp. 69-92 in Population-Level Ecological Risk Assessment, L.W. Barnthouse, W.R. Munns, and M.T. Sorensen, eds. Boca Raton: CRC Press.
Moore, D.R.J., W.J. Warren-Hicks, S. Qian, A. Fairbrother, T. Aldenberg, T. Barry, R. Luttik, and H.T. Ratte. 2010. Uncertainty analysis using classical and Bayesian hierarchical models. Pp. 123-141 in Application of Uncertainty Analysis to Ecological Risks of Pesticides, W.J. Warren-Hicks, and A. Hart, eds. Pensacola, FL: SETAC Press.
Morris, W.F., and D.F. Doak. 2002. Quantitative Conservation Biology: Theory and Practice of Population Viability Analysis. Sunderland, MA: Sinauer Associates.
Mu, X., and G.A. LeBlanc. 2004. Synergistic interaction of endocrine disrupting chemicals: Model development using an ecdysone receptor antagonist and a hormone synthesis inhibitor. Environ. Toxicol. Chem. 23(4):1085-1091.
Munns, W.R., D.E. Black, T.R. Gleason, K. Salomon, D. Bengtson, and R. Gutjahr-Gobell. 1997. Evaluation of the effects of dioxin and PCBs on *Fundulus heteroclitus* populations using a modeling approach. Environ. Toxicol. Chem. 16(5):1074-1081.
Murphy, S.D., R.L. Anderson, and K.P. DuBois. 1959. Potentiation of toxicity of malathion by triorthotolyl phosphate. Proc. Soc. Exp. Biol. Med. 100(3):483-487.
Newman, M.C., D.R. Ownby, L.C.A. Mézin, D.C. Powell, T.R.L. Christensen, S.B. Lerberg, and B.A. Anderson. 2000. Applying species-sensitivity distributions in ecological risk assessment: Assumptions of distribution type and sufficient numbers of species. Environ. Toxicol. Chem. 19 (2):508-515.
Niederlehner, B.R., J. Cairns, Jr., and E.P. Smith. 1998. Modeling acute and chronic toxicity of nonpolar narcotic chemicals and mixtures to *Ceriodaphnia dubia*. Ecotoxicol. Environ. Saf. 39(2):136-146.
Nielsen, J.L. 1992. Microhabitat-specific foraging behavior, diet, and growth of juvenile coho salmon. T. Am. Fish. Soc. 121(5):617-634.
NMFS (National Marine Fisheries Service). 2008. Biological Opinion, Environmental Protection Agency Registration of Pesticides Containing Chlorpyrifos, Diazinon, and Malathion. National Marine Fisheries Service, Silver Spring, MD. November 18, 2008 [online]. Available: http://www.nmfs.noaa.gov/pr/pdfs/pesticide_biop.pdf [accessed Nov. 13, 2012].
NMFS (National Marine Fisheries Service). 2009. Biological Opinion, Endangered Species Act Section 7 Consultation, Environmental Protection Agency Registration of Pesticides Containing Carbaryl, Carbofuran, and Methomyl. National Marine Fisheries Service, Silver Spring, MD. April 20, 2009 [online]. Available: http://www.nmfs. noaa.gov/pr/pdfs/carbamate.pdf [accessed Nov. 12, 2012].
NMFS (National Marine Fisheries Service). 2010. Biological Opinion, Endangered Species Act Section 7 Consultation, Environmental Protection Agency Registration of Pesticides Containing Azinphos methyl, Bensulide, Dimethoate, Disulfoton, Ethoprop, Fenamiphos, Naled, Methamidophos, Methidathion, Methyl parathion, Phorate and Phosmet. National Marine Fisheries Service, Silver Spring, MD. August 31, 2010 [online]. Available: http://www.nmfs.noaa.gov/pr/pdfs/final_batch_3_ opinion.pdf [accessed Nov. 13, 2012].
NMFS (National Marine Fisheries Service). 2011. Biological Opinion, Endangered Species Act Section 7 Consultation, Environmental Protection Agency Registration of Pesticides 2,4-D, Triclopyr BEE, Diuron, Linuron, Captan, and Chlorothalonil. National Marine Fisheries Service, Silver Spring, MD. June 30, 2011[online]. Available: http://www.nmfs.noaa.gov/pr/pdfs/consultations/pesticide_opinion4.pdf [accessed Nov. 13, 2012].

NMFS (National Marine Fishers Service). 2012. NOAA Fisheries (NMFS) Response to Questions from the Committee on Ecological Risk Assessment under FIFRA and ESA, from Scott Hecht, NOAA, to David Policansky, March 30, 2012 [online]. Available: http://www.thecre.com/forum1/wp-content/uploads/2012/04/esanmfsatrazine.pdf [accessed Nov. 19, 2012].

Norgaard, K.B., and N. Cedergreen. 2010. Pesticide cocktails can interact synergistically on aquatic crustaceans. Environ. Sci. Pollut. Res. 17(4):957-967.

NRC (National Research Council). 1996. Upstream: Salmon and Society and Pacific Northwest. Washington, DC: National Academy Press.

NRC (National Research Council). 2003. Cumulative Environmental Effects of Oil and Gas Activities on Alaska's North Slope. Washington, DC: National Academies Press.

OECD (Organisation for Economic Co-operation and Development). 2001. Harmonized Integrated Classification System for Human Health and Environmental Hazards of Chemical Substances and Mixtures. ENV/JM/MONO(2001)6. OECD Series on Testing and Assessment Number 33. OECD [online]. Available: http://iccvam.niehs.nih.gov/SuppDocs/FedDocs/OECD/OECD_hclfinaw.pdf [accessed Nov. 19, 2012].

Oliveira, A.G., L.F. Telles, R.A. Hess, G.A. Mahecha, and C.A. Oliveira. 2007. Effects of the herbicide Roundup on the epididymal region of drakes, *Anas platyrhynchos*. Reprod. Toxicol. 23(2):182-191.

Olmstead, A.W., and G.A. LeBlanc. 2005a. Toxicity assessment of environmentally-relevant pollutant mixtures using a heuristic model. Integr. Environ. Assessm. Manag. 1(2):114-122.

Olmstead, A.W., and G.A. LeBlanc. 2005b. Joint action of polycyclic aromatic hydrocarbons: Predictive modeling of sublethal toxicity. Aquat. Toxicol. 75(3):253-262.

Orlando, J.L., K.M. Kuivila, and A. Whitehead. 2003. Dissolved Pesticide Concentrations Detected in Storm-water Runoff at Selected Sites in the San Joaquin River Basin, California, 2000-2001. Open-File Report 03-101. U.S. Geological Survey, Reston, VA [online]. Available: http://pubs.usgs.gov/of/2003/ofr03101/pdf/ofr03101.pdf [accessed Nov. 19, 2012].

Orlando, J.L., L.A. Jacobson, and K.M. Kuivila. 2004. Dissolved Pesticide and Organic Carbon Concentrations Detected in Surface Waters, Northern Central Valley, California, 2001-2002. Open File Report 2004-1214. U.S. Geological Survey, Reston, VA [online]. Available: http://pubs.usgs.gov/of/2004/1214/ofr2004-1214.pdf [accessed Nov. 19, 2012].

Ortiz, D., L. Yanez, H. Gomez, J.A. Martinez-Salazar, and F. Diaz-Barriga. 1995. Acute toxicological effects in rats treated with a mixture of commercially formulated products containing methyl parathion and permethrin. Ecotoxicol. Environ. Saf. 32(2):154-158.

Palmqvist, A., and V.E. Forbes. 2008. Demographic effects of the polycyclic aromatic hydrocarbon, fluoranthene, on two sibling species of the Polychaete *Capitella capitata*. Pp. 200-212 in Demographic Toxicity: Methods in Ecological Risk Assessment, H.R. Akçakaya, J.D. Stark, and T.S. Bridges, eds. New York: Oxford University Press.

Pastorok, R.A., S.M. Bartell, S. Ferson, and L.R. Ginzburg, eds. 2002. Ecological Modeling in Risk Assessment: Chemical Effects on Populations, Ecosystems, and Landscapes. Boca Raton, FL: CRC Press.

Posthuma, L., G.W. Suter, and T.P. Traas, eds. 2001. Species Sensitivity Distributions in Ecotoxicology. Boca Raton, FL: CRC Press.

Puccia, C.J., and R. Levins. 1991. Qualitative modeling in ecology: Loop analysis, signed digraphs, and time averaging. Pp. 119-143 in Qualitative Simulation Modeling and Analysis, P.A. Fishwick, and P.A. Luker, eds. New York: Springer.

Quinn, T.P. 2005. The Behavior and Ecology of Pacific Salmon and Trout. Seattle: University of Washington Press.

Quinn, T.J., II, and R.B. Deriso. 1999. Quantitative Fish Dynamics. New York: Oxford University Press.

Raimondo, S., D.N. Vivian, C. Delos, and M.G. Barron. 2008. Protectiveness of species sensitivity distribution hazard concentrations for acute toxicity used in endangered species risk assessment. Environ. Toxicol. Chem. 27(12):2599-2607.

Raimondo, S., C.R. Jackson, and M.G. Barron. 2010. Influence of taxonomic relatedness and chemical mode of action in acute interspecies estimation models for aquatic species. Environ. Sci. Technol. 44(19):7711-7716.

RAMAS. 2011. Incorrectly modeling impact under density dependence. Avoiding Mistakes in Population Modeling: Density Dependence [online]. Available: http://www.ramas.com/CMdd.htm#ddimpact [accessed Apr. 18, 2013].

Rider, C.V., and G.A. LeBlanc. 2005. An integrated addition and interaction model for assessing toxicity of chemical mixtures. Toxicol. Sci. 87(2):520-528.

Rohr, J.R., J.L. Kerby, and A. Sih. 2006. Community ecology as a framework for predicting contaminant effects. Trends Ecol. Evol. 21(11):606-613.

Safe, S. 1998. Hazard and risk assessment of chemical mixtures using the toxic equivalency factor approach. Environ. Health Perspect. 106(suppl. 4):1051-1058.

Sahay, N., and R.A. Agarwal. 1997. MGK-264-Pyrethroid synergism against *Lymnaea acuminata*. Chemosphere 35(5):1011-1021.

Sandahl, J.F., D.H. Baldwin, J.J. Jenkins, and N.L. Scholz. 2005. Comparative thresholds for acetylcholinesterase inhibition and behavioral impairment in coho salmon exposed to chlorpyrifos. Environ. Toxicol. Chem. 25(1):136-145.

Sappington, L.C., F.L. Mayer, F.J. Dwyer, D.R. Buckler, J.R. Jones, and M.R. Ellersieck. 2001. Contaminant sensitivity of threatened and endangered fishes compared to standard surrogate species. Environ. Toxicol. Chem. 20(12):2869-2876.

Schleier, J.J., and R.K. Peterson. 2012. The joint toxicity of type I, II, and nonester pyrethroid insecticides. J. Econ. Entomol. 105(1):85-91.

Scholz, N.L., N.K. Truelove, B.L. French, B.A. Berejikian, T.P. Quinn, E. Casillas, and T.K. Collier. 2000. Diazinon disrupts antipredator and homing behavior in chinook salmon (*Oncorhynchus tshawytscha*). Can. J. Fish. Aquat. Sci. 57(9):1911-1918.

Scholz, N.L., N.K. Truelove, J.S. Labenia, D.H. Baldwin, and T.K. Collier. 2006. Dose-additive inhibition of chinook salmon acetylcholinesterase activity by mixtures of organophosphate and carbamate insecticides. Environ. Toxicol. Chem. 25(5):1200-1207.

Smyth, H.F., C.S. Weil, J.S. West, and C.P. Carpenter. 1969. An exploration of joint toxic action: Twenty-seven industrial chemicals intubated in rats in all possible pairs. Toxicol. Appl. Pharmacol. 14(2):340-347.

Solomon, K.R., T.C.M. Brock, D. de Zwart, S.D. Dyer, L. Posthuma, S.M. Richards, H. Sanderson, P.K. Sibley, and P.J. van den Brink. 2008. Extrapolation Practice for Ecotoxicological Effect Characterization of Chemicals. Pensacola, FL: SETAC Press. 380pp.

Spear, R.C. 1997. Large simulation models: Calibration, uniqueness and goodness of fit. Environ. Modell. Softw. 12(2-3):219-228.

Spencer, K.A., K.L. Buchanan, A.R. Goldsmith, and C.K. Catchpole. 2003. Song as an honest signal of developmental stress in zebra finch (*Taeniopygia guttata*). Horm. Behav. 44(2):132-139.

Spromberg, J.A., and L.L. Johnson. 2008. Potential effects of freshwater and estuarine contaminant exposure on lower Columbia River Chinook salmon (*Oncorhynchus tshawytscha*) populations. Pp. 123-142 in Demographic Toxicity: Methods in Ecological Risk Assessment, H.R. Akçakaya, J.D. Stark, and T.S. Bridges, eds. New York: Oxford University Press.

Stark, J.D. 2012. Demography and modeling to improve pesticide risk assessment of endangered species. Pp. 259-270 in Pesticide Regulation and the Endangered Species Act, K.D. Racke, B.D. McGaughey, J.L. Cowles, A.T. Hall, S.H. Jackson, J.J. Jenkins, J.J. Johnston, eds. ACS Symposium Series 111. Washington, DC: American Chemical Society.

Suttle, K.B., M.E. Power, J.M. Levine, and C. McNeely. 2004. How fine sediment in riverbeds impairs growth and survival of juvenile salmonids. Ecol. Appl. 14(4):969-974.

Taylor, E.B. 1990. Environmental correlates of life-history variation in juvenile chinook salmon, *Oncorhynchus tshawytscha* (Walbaum). J. Fish Biol. 37(1):1-17.

TenBrook, P.L., A.J. Palumbo, T.L. Fojut, P. Hann, J. Karkoski, and R.S. Tjeerdema. 2010. The University of California-Davis methodology for deriving aquatic life pesticide water quality criteria. Rev. Environ. Contam. Toxicol. 209:1-155.

Thompson, H.M. 1996. Interactions between pesticides: A review of reported effects and their implications for wildlife risk assessment. Ecotoxicology 5(2):59-81.

Topping, C.J., R.M. Sibly, H.R. Akçakaya, G.C. Smith, and D.R. Crocker. 2005. Risk assessment of UK skylark populations using life-history and individual-based landscape models. Ecotoxicology 14(8):925-936.

Verhoef, H.A., and P.J. Morin. 2010. Community Ecology: Processes, Models, and Applications. New York: Oxford University Press. 247pp.

Verma, S.R., S. Rani, S.K. Bansal, and R.C. Dalela. 1980. Effects of the pesticides thiotox, dichlorvos and carbofuran on the test fish *Mystus vittatus*. Water Air Soil Pollut. 13(2):229-234.

Walter, H., F. Consolaro, P. Gramatica, M. Scholze, and R. Altenburger. 2002. Mixture toxicity of priority pollutants at No Observed Effect Concentrations (NOECs). Ecotoxicology 11(5):299-310.

Warren-Hicks, W.J., and A Hart, eds. 2010. Application of Uncertainty Analysis to Ecological Risks of Pesticides. Pensacola, FL: SETAC Press.

Werner, I., L.A. Deanovic, D. Markiewicz, M. Khamphanh, C.K. Reece, M. Stillway, and C. Reece. 2010. Monitoring acute and chronic water column toxicity in the northern Sacramento-San Joaquin estuary, California, USA, using the euryhaline amphipod, *Hyalella azteca*: 2006 to 2007. Environ. Toxicol. Chem. 29(10):2190-2199.

WHO (World Health Organization). 1994. Glyphosate. Environmental Health Criteria 159. Geneva: World Health Organization [online]. Available: http://www.inchem.org/documents/ehc/ehc/ehc159.htm [accessed Feb. 19, 2013].

Williams, G.M., R. Kroes, and I.C. Munro. 2000. Safety evaluation and risk assessment of the herbicide Roundup and its active ingredient, glyphosate, for humans. Regul. Toxicol. Pharmacol. 31(2 Pt 1):117-165.

Willson, J.D., W.A. Hopkins, C.M. Bergeron, and B.D. Todd. 2012. Making leaps in amphibian ecotoxicology: Translating individual-level effects of contaminants to population viability. Ecol. Appl. 22(6):1791-1802.

Yang, R.S., and J.E. Dennison. 2007. Initial analyses of the relationship between "Thresholds" of toxicity for individual chemicals and "Interaction Thresholds" for chemical mixtures. Toxicol. Appl. Pharmacol. 223(2):133-138.

5

Risk Characterization

Risk characterization is the final stage of an ecological risk assessment in which results of exposure and effects analyses are integrated to provide decision-makers with a risk estimate—the probability of adverse effects of exposure to a chemical stressor—and its associated uncertainty. A decision-maker does not want to make a decision on the basis of a belief that a pesticide is unlikely to yield an adverse effect and discover afterwards that it did yield an adverse effect. That is often referred to as avoiding a Type II error. For example, if the US Environmental Protection Agency (EPA) proposes registering a pesticide with a specific label use, it needs to know how much confidence there is that doing so will lead to the desired outcome, such as reduction in the abundance of the target species, and not result in jeopardy to a listed species. It is most useful if the risk estimate and its associated uncertainty are expressed in a quantitative manner—for example, "there is a 20% ± 10% probability of a 25% reduction in the population growth rate as a result of this action."

In addition to generating a quantitative risk estimate, risk characterization includes a narrative discussion (termed the risk description) that includes discussion of data gaps, lack of knowledge, natural variability, and other factors that might influence confidence in the risk estimate. The discussion can be viewed as a weight-of-evidence description in which the strengths and weaknesses of each assumption and each type of data used in the risk assessment are discussed. At the risk assessor's discretion, the narrative might be summarized in a table that lists all the lines of evidence and their various weights that are scored on the basis of relevance, degree of quantification, variability, and robustness of the data analysis (see, for example, Linkov et al. 2009; Exponent 2010). The discussion provides guidance to the decision-maker about which aspects of the risk assessment are more reliable, where there are greater unknowns, and how natural variability or lack of knowledge might hinder the development of a more accurate estimate of risk.

There are many practical methods for combining the results (with their associated uncertainties) of exposure and effects analyses to provide an estimate of risk and the confidence in it. Two broad approaches have been used; one is a deterministic concentration-ratio approach, which compares point estimates of exposure and effect concentrations, and the other is a probabilistic approach, which evaluates the probability that exposure to a chemical will lead to a specified adverse effect at some future time. The latter is technically sound, and the former is ad hoc (although commonly used) and has unpredictable performance outcomes. EPA uses the concentration-ratio approach for its assessments. In biological opinions on salmon, the National Marine Fisheries Service appears to favor a probabilistic approach that is based on population modeling. The Fish and Wildlife Service seems not to use a quantitative approach, either concentration-ratio or probabilistic, for risk characterization.

CONCENTRATION-RATIO APPROACH

The concentration-ratio approach, which is commonly used by EPA for Step 1 and 2 assessments (see Figure 2-1), does not estimate risk (the probability of an adverse effect) itself but rather relies on there being a large margin between a point estimate of the most likely maximum pesticide environmental concentration and a point estimate of the lowest concentration at which a specified adverse effect might be expected (EPA 2004). The superficial attraction of this approach is that one feels confident that a decision will not lead to an adverse effect (that is, a Type II error will be avoided) if sufficiently large margins are used. There is a belief that the larger the margin between the estimated exposure and the response threshold, the greater the certainty (or the smaller the uncertainty). The flaws in that approach are that it does not account for the probability of an adverse effect before worst-case assumptions are applied and that it does not calculate how the use of the assumptions modifies that probability. Given that approach, decision-makers do not know what the probability of an adverse effect is, but they hope that they can assume (or be reassured) that it is small. However, such an assumption is not reliable. If they or their constituencies have doubts, the common response is to widen the margin with additional conservative assumptions, including addition of specific uncertainty factors or more stringent, and possibly implausible, exposure scenarios. However, simply widening the gap indefinitely might lead to decisions that limit pesticide use to a greater extent than is intended by policy and will not meaningfully express the underlying probability of an adverse effect.

For pesticides, as evaluated by EPA, the concentration ratio is quantified in the form of a risk quotient (RQ) that might be less or greater than some specified level of concern (LOC). However, an RQ is not actually a risk estimate in that it provides no information about the probability of an adverse effect. Thus, although an RQ of 10 is several times higher than most numerical LOCs, there is no fixed relationship between RQs and the probability of an adverse effect on a listed species. Therefore, it is not possible to determine what an RQ of 10 means with respect to a possible adverse effect on a listed species. Nor is there a fixed relationship for comparing the difference between, for example, RQs of 10 and

100 with respect to the probability of an adverse effect. Theoretically, an RQ of 100 means a greater probability of an adverse effect than an RQ of 10, but one cannot determine whether the difference in probability between the two RQs is substantial or negligible or whether the final error associated with the risk estimate is appropriate for the management needs.

Thus, although RQs are often used by EPA for Step 2 assessments that might trigger later, more refined and focused assessments for listed species, the committee concludes that RQs are not appropriate for assessments for listed species or indeed for any application in which it is desired to base a decision on the probabilities of various possible outcomes. Furthermore, the committee concludes that adding uncertainty factors to RQs to account for lack of data (on formulation toxicity, synergy, additivity, or any other aspect) is unwarranted because there is no way to determine whether the assumptions being used substantially overestimate or underestimate the probability of an adverse effect.

The committee has not been asked about and is not commenting on policy decisions about what level of risk is acceptable or how conservative the agencies should be in establishing an "acceptable" risk level when considering jeopardy to listed species.

PROBABILISTIC APPROACH

Risk is defined as the probability of an adverse effect (Burmaster 1996). Thus, natural tools for quantifying and analyzing risk are probability, statistics, and the algebra of random variables, and an alternative to the deterministic concentration-ratio approach is a probabilistic one. In the probabilistic approach, the probability that a decision will lead to an adverse effect is calculated from the available information and then used to support an informed decision (again, the committee is purposefully refraining from a discussion of what an "acceptable" probability of risk might be). The probabilistic approach requires integration of the uncertainties (from sampling, natural variability, lack of knowledge, and measurement and model error) in the exposure and effects analyses by using probability distributions, rather than single point estimates, for uncertain quantities (EPA 2001). The distributions are then integrated mathematically to calculate the risk as a probability and its associated uncertainty in that estimate. Ultimately, decision-makers are provided with a risk estimate that reflects the probability of exposure to a range of pesticide concentrations and the magnitude of an adverse effect (if any) to the exposures that answers the fundamental question, What is the probability that registration of this pesticide will lead to a specified adverse effect on a listed species or its critical habitat?

Implementing a probabilistic approach requires three primary actions on the part of a risk assessor:

(1) Describe uncertain variables with distributions and recognize that not all variables in a model or an analysis need be treated this way. The task can be

made considerably more tractable if only variables identified as key drivers via a sensitivity analysis are defined by distributions. The methods and problems in fitting or otherwise deriving the distributions from data are not discussed here because a large literature is available on these topics (see, for example, EUFRAM 2006; Warren-Hicks and Hart 2010). However, the models or measurements used to estimate exposure concentrations are capable of providing results as distributions, and results of the multispecies toxicity testing that is already part of the registration process could be expressed as discrete exposure-response distributions or combined into a species sensitivity distribution.

(2) Propagate the uncertainty through to distributions of exposure and effect by using one of several calculation methods. The most readily accessible of these (in terms of software and experience) are Monte Carlo analysis (including second-order methods), probability-bounds analysis, and Bayesian methods (Warren-Hicks and Hart 2010) (see Chapter 2 for recommendations of method selection).

(3) Integrate exposure and effect estimates to calculate risk. Aldenberg et al. (2001) have shown that a variety of risk-estimation methods calculate the same probability that a stated exposure concentration will produce a specified adverse effect given a specific exposure-response relationship. Such methods include discrete summation for expected risk (Cardwell et al. 1999), ecological risk overlap plot (Van Straalen 2002), numerical integration of risk-distribution curves (Parkhurst et al. 1996; Solomon and Takacs 2001; Warren-Hicks et al. 2001), and various area under the curve (AUC) methods, such as exceedance profile plots (ECOFRAM 1999ab; Giesy et al. 1999; Solomon and Takacs 2001), cumulative profile plots (Aldenberg et al. 2001), and cumulative distribution functions of risk estimates (Aldenberg et al. 2001; EUFRAM 2006). The area under the joint probability curve is considered as a numerical measure of the risk to a species posed by a chemical stressor (Giddings et al. 2005), a value that a decision-maker would seek to minimize.

The committee has concluded that EPA and the Services can begin the transition now from concentration ratios to established, scientifically defensible statistical-inference methods for propagating uncertainties in exposure and effect through to a risk estimate for both individual receptors (Step 2) and populations of receptors (Step 3). The committee recognizes the pragmatic demands of the pesticide registration process and encourages EPA and the Services to consider probabilistic methods that have already been successfully applied to pesticide risk assessments (Odenkirchen 2003 [EPA's Terrestrial Investigation Model v 2.0]; Giddings et al 2005; Warren-Hicks and Hart 2010), have otherwise appeared frequently in the technical literature, are familiar to many risk-assessment practitioners, can be implemented with commercially available software, and are most readily explicable to decision-makers, stakeholders, and the public. The committee also notes that transitioning to a probabilistic approach can begin with simple registrations (for example, pesticides for use on a few crops or in a small geographic area) and will not require that all variables be immediately represented with probability distributions (that is, sensitivity analyses can be used to identify key parameters that are important to represent as probability distributions).

CONCLUSIONS

- Inclusion of uncertainty factors to account for lack of various data is unwarranted because there is no way to determine whether the assumptions being used substantially overestimate or underestimate the probability of adverse effects.
- RQs are not appropriate for risk assessments or for any application in which it is desired to base a decision on the probabilities of the various possible outcomes.
- EPA (for Step 2 assessments) and the Services (for Step 3 assessments) should use established, scientifically defensible, statistical methods to calculate risk as a probability to assist decision-makers' understanding of the potential consequences of their decisions.
- A number of existing probabilistic methods have been shown to be applicable and practical for ecological risk assessments that involve pesticides.
- The transition from concentration-ratio to probabilistic approaches can begin now, starting with simple registrations, focusing on a small set of sensitive key parameters, and drawing on the considerable literature and guidance on probabilistic approaches.

REFERENCES

Aldenberg, T., J.S. Jaworska, and T.P. Traas. 2001. Normal species sensitivity distributions and probabilistic ecological risk assessment. Pp. 49-102 in Species Sensitivity Distributions in Ecotoxicology, L. Posthuma, G.W. Suter, II, and T.P. Traas, eds. Boca Raton: CRC Press.

Burmaster, D.E. 1996. Benefits and costs of using probabilistic techniques in human health risk assessments—with an emphasis on site-specific risk assessments. Hum. Ecol. Risk Assess. 2(1):35-43.

Cardwell, R.D., M.S. Brancato, J. Toll, D. DeForest, and L. Tear. 1999. Aquatic ecological risks posed by tributyltin in United States surface waters: Pre-1989 to 1996 data. Environ. Toxicol. Chem. 18(3):567-577.

ECOFRAM (Ecological Committee on FIFRA Risk Assessment Methods).1999a. ECOFRAM Aquatic Report. Office of Pesticide Programs, U.S. Environmental Protection Agency, Washington, DC. May 4, 1999 [online]. Available: http://www.epa.gov/oppefed1/ecorisk/aquareport.pdf [accessed Nov. 21, 2012].

ECOFRAM (Ecological Committee on FIFRA Risk Assessment Methods). 1999b. ECOFRAM Terrestrial Draft Report. Office of Pesticide Programs, U.S. Environmental Protection Agency, Washington, DC. May 10, 1999 [online]. Available: http://www.epa.gov/oppefed1/ecorisk/terrreport.pdf [accessed Nov. 21, 2012].

EPA (U.S. Environmental Protection Agency). 2001. Risk Assessment Guidance for Superfund: Volume III - Part A. Process for Conducting Probabilistic Risk Assessment. EPA 540-R-02-002, OSWER 9285.7-45. Office of Emergency and Remedial Response, U.S. Environmental Protection Agency, Washington, DC [online]. Available: http://www.epa.gov/oswer/riskassessment/rags3adt/pdf/rags3 adt_complete.pdf [accessed Nov. 21, 2012].

EPA (U.S. Environmental Protection Agency). 2004. Overview of the Ecological Risk Assessment Process in the Office of Pesticide Programs, U.S. Environmental Protection Agency: Endangered and Threatened Species Effects Determinations. Office of Prevention, Pesticides and Toxic Substances, Office of Pesticide Programs, U.S. Environmental Protection Agency, Washington DC. January 23, 2004 [online]. Available: http://www.epa.gov/oppfead1/endanger/con sultation/ecorisk-overview.pdf [accessed Aug. 29, 2012].

EUFRAM. 2006. Concerted Action to Develop a European Framework for Probabilistic Risk Assessment of the Environmental Impacts of Pesticides. Fifth Framework Programme, Quality of Life and Management of Living Resources, European Commission, Brussels [online]. Available: http://www.eufram.com/ [accessed Nov. 21, 2012].

Exponent. 2010. Report on Guidance for a Weight of Evidence Approach in Conducting Detailed Ecological Risk Assessments (DERA) in British Columbia. Prepared for The Ministry of Environment, British Columbia, by Exponent, Inc., Bellevue, WA. June 2010 [online]. Available: http://www.sabcs.chem.uvic.ca/woe. html [accessed Aug. 2012].

Giddings, J.M., T.A. Anderson, L.W. Hall, Jr., A.J. Hosmer, R.J. Kendall, R.P. Richards, K.R. Solomon, and W.M. Williams. 2005. Atrazine in North American Surface Waters: A Probabilistic Aquatic Ecological Risk Assessment. Pensacola, FL: SETAC Press. 392 pp.

Giesy, J.P., K.R. Solomon, J.R. Coats, K.R. Dixon, J.M. Giddings, and E.E. Kenaga. 1999. Chlorpyrifos: Ecological risk assessment in North American aquatic environments. Rev. Environ. Contam. Toxicol. 160:1-129.

Linkov, I., D. Loney, S. Cormier, K. Satterstrom, and T. Bridges. 2009. Weight-of-evidence evaluation in environmental assessment: Review of qualitative and quantitative approaches. Sci. Total Environ. 407(19):5199-5205.

Odenkirchen, E. 2003. Evolution of OPP's Terrestrial Investigation Model Software and Programming to Meet Technical/Regulatory Challenges. Presentation at EUropean FRamework for Probabilistic Risk Assessment of the Environmental Impacts of Pesticides Workshop, June 5-8, 2003. Bilthoven, the Netherlands [online]. Available: http://www.epa.gov/oppefed1/ecorisk/presentations/eufram _overview.htm [accessed Mar. 28, 2013].

Parkhurst, B.R., W. Warren-Hicks, T. Etchison, J.B. Butcher, R.D. Cardwell and J. Voloson. 1996. Methodology for Aquatic Ecological Risk Assessment. RP91-AER-1 1995. Water Environment Research Foundation, Alexandria, VA.

Solomon, K.R., and P. Takacs. 2001. Probabilistic risk assessment using species sensitivity distributions. Pp. 285-314 in Species Sensitivity Distributions in Ecotoxicology, L. Posthuma, G.W. Suter, II, and T.P. Traas, eds. Boca Raton: CRC Press.

Van Straalen, N.M. 2002. Threshold models for species sensitivity distributions applied to aquatic risk assessment for zinc. Environ. Toxicol. Pharmacol. 11(3-4):167-172.

Warren-Hicks, W.J., and A. Hart, eds. 2010. Application of Uncertainty Analysis to Ecological Risks of Pesticides. Pensacola, FL: SETAC Press. 228 pp.

Warren-Hicks, W.J., B.R. Parkhurst, and J.B. Butcher. 2001. Methodology for aquatic ecological risk assessment. Pp. 345-382 in Species Sensitivity Distributions in Ecotoxicology, L. Posthuma, G.W. Suter, II, and T.P. Traas, eds. Boca Raton: CRC Press.

Appendix A

Selected Excerpts from 40 CFR Part 158 Data Requirements for Pesticides

158.1 Purpose and scope.

a) *Purpose.* The purpose of this part is to specify the kinds of data and information EPA requires in order to make regulatory judgments under FIFRA secs. 3, 4, and 5 about the risks and benefits of pesticide products. Further, this part specifies the data and information needed to determine the safety of pesticide chemical residues under FFDCA sec. 408.

b) *Scope*—

(1) This part describes the minimum data and information EPA typically requires to support an application for pesticide registration or amendment; support the reregistration of a pesticide product; support the maintenance of a pesticide registration by means of the data call-in process, e.g., as used in the registration review program; or establish or maintain a tolerance or exemption from the requirements of a tolerance for a pesticide chemical residue.

(2) This part establishes general policies and procedures associated with the submission of data in support of a pesticide regulatory action.

(3) This part does not include study protocols, methodology, or standards for conducting or reporting test results; nor does this part describe how the Agency uses or evaluates the data and information in its risk assessment and risk management decisions, or the regulatory determinations that may be based upon the data.

* * *

§ 158.30 Flexibility.

a) FIFRA provides EPA flexibility to require, or not require, data and information for the purposes of making regulatory judgments for pesticide products. EPA has the authority to establish or modify data needs for individual pesticide chemicals. The actual data required may be modified on an individual basis to fully characterize the use and properties, characteristics, or effects of specific pesticide products under review. The Agency encourages each applicant to consult with EPA to discuss the data requirements particular to its product prior to and during the registration process.

b) The Agency cautions applicants that the data routinely required in this part may not be sufficient to permit EPA to evaluate the potential of the product to cause unreasonable adverse effects to man or the environment. EPA may require the submission of additional data or information beyond that specified in this part if such data or information are needed to appropriately evaluate a pesticide product.

c) This part will be updated as needed to reflect evolving program needs and advances in science.

* * *

§ 158.70 Satisfying data requirements.

a) *General policy.* The Agency will determine whether the data submitted or cited to fulfill the data requirements specified in this part are acceptable. This determination will be based on the design and conduct of the experiment from which the data were derived, and an evaluation of whether the data fulfill the purpose(s) of the data requirement. In evaluating experimental design, the Agency will consider whether generally accepted methods were used, sufficient numbers of measurements were made to achieve statistical reliability, and sufficient controls were built into all phases of the experiment. The Agency will evaluate the conduct of each experiment in terms of whether the study was conducted in conformance with the design, good laboratory practices were observed, and results were reproducible. The Agency will not reject data merely because they were derived from studies which, when initiated, were in accordance with an Agency-recommended protocol, even if the Agency subsequently recommends a different protocol, as long as the data fulfill the purposes of the requirements as described in this paragraph.

(1) The provisions in this part 158 should be read in conjunction with the provisions in §152.85 to claim eligibility for the formulators' exemption.

(2) [Reserved]

b) *Good laboratory practices.* Applicants must adhere to the good laboratory practice (GLP) standards described in 40 CFR part 160 when conducting studies. Applicants must also adhere to GLP standards when conducting a study in support of a waiver request of any data requirement which is within the scope of the GLP requirements.

c) *Agency guidelines.* EPA has published Test Guidelines that contain standards for conducting acceptable tests, guidance on the evaluation and reporting of data, definition of terms, and suggested study protocols. Copies of the Test Guidelines may be obtained by visiting the agency's website at *www.epa.gov/pesticides*.

Appendix A

d) *Study protocols—*

(1) *General.* Any appropriate protocol may be used to generate the data required by this part, provided that it meets the purpose of the test standards specified in the pesticide assessment guidelines, and provides data of suitable quality and completeness as typified by the protocols cited in the guidelines. Applicants should use the test procedure which is most suitable for evaluation of the particular ingredient, mixture, or product. Accordingly, failure to follow a suggested protocol will not invalidate a test if another appropriate methodology is used.

(2) *Organization for Economic Co-Operation and Development (OECD) protocols.* Tests conducted in accordance with the requirements and recommendations of the applicable OECD protocols can be used to develop data necessary to meet the requirements specified in this part. Applicants should note, however, that certain of the OECD recommended test standards, such as test duration and selection of test species, are less restrictive than those recommended by EPA. Therefore, when using OECD protocols, care should be taken to observe the test standards in a manner such that the data generated by the study will satisfy the requirements of this part.

e) *Combining studies.* Certain toxicology studies may be combined to satisfy data requirements. For example, carcinogenicity studies in rats may be combined with the rat chronic toxicity study. Combining appropriate studies may be expected to reduce usage of test animals as well as reduce the cost of studies. EPA encourages this practice by including standards for acceptable combined tests in the Pesticide Assessment Guidelines. Registrants and applicants are encouraged to consider combining other tests when practical and likely to produce scientifically acceptable results. Registrants and applicants, however, must consult with the EPA before initiating combined studies.

* * *

§ 158.75 Requirements for additional data.

The data routinely required by this part may not be sufficient to permit EPA to evaluate every pesticide product. If the information required under this part is not sufficient to evaluate the potential of the product to cause unreasonable adverse effects on man or the environment, additional data requirements will be imposed. However, EPA expects that the information required by this part will be adequate in most cases for an assessment of the properties and effects of the pesticide.

§ 158.80 Use of other data.

a) *Data developed in foreign countries.* With certain exceptions, laboratory and field study data developed outside the United States may be submitted in support of a pesticide registration. Data generated in a foreign country which the

Agency will not consider include, but are not limited to, data from tests which involved field test sites or a test material, such as a native soil, plant, or animal, that is not characteristic of the United States. Applicants submitting foreign data must take steps to ensure that U.S. materials are used, or be prepared to supply data or information to demonstrate the lack of substantial or relevant differences between the selected material or test site and the U.S. material or test site. Once submitted, the Agency will determine whether or not the data meet the data requirements.

b) *Data generated for other purposes.* Data developed for purposes other than satisfaction of FIFRA data requirements, such as monitoring studies, may also satisfy data requirements in this part. Consultation with the Agency should be arranged if applicants are unsure about suitability of such data.

* * *

§ 158.130 Purposes of the registration data requirements.

e) *Hazards to nontarget organisms*—

(1) *General.* The information required to assess hazards to nontarget organisms is derived from tests to determine pesticidal effects on birds, mammals, fish, terrestrial and aquatic invertebrates and plants. These tests include short-term acute, subacute, reproduction, simulated field, and full field studies arranged in a hierarchical or tier system which progresses from the basic laboratory tests to the applied field tests. The results of each tier of testing must be evaluated to determine the potential of the pesticide to cause adverse effects, and to determine whether further testing is required. A purpose common to all data requirements is to provide data which determine the need for (and appropriate wording for) precautionary label statements to minimize the potential adverse effects to nontarget organisms.

(2) *Short-term studies.* The short-term acute and subchronic laboratory studies provide basic toxicity information which serves as a starting point for the hazard assessment. These data are used: To establish acute toxicity levels of the active ingredient to the test organisms; to compare toxicity information with measured or estimated pesticide residues in the environment in order to assess potential impacts on fish, wildlife and other nontarget organisms; and to indicate whether further laboratory and/or field studies are needed.

(3) *Long-term and field studies.* Additional studies (*i.e.*, avian, fish, and invertebrate reproduction, life cycle studies and plant field studies) may be required when basic data and environmental conditions suggest possible problems. Data from these studies are used to: Estimate the potential for chronic effects, taking into account the measured or estimated residues in the environment; and to determine if additional field or laboratory data are necessary to further evaluate hazards. Simulated field and/or field data are used to examine acute and chronic adverse effects

Appendix A

on captive or monitored fish and wildlife populations under natural or near-natural environments. Such studies are required only when predictions as to possible adverse effects in less extensive studies cannot be made, or when the potential for adverse effects is high.

* * *

g) *Pesticide spray-drift evaluation.* Data required to evaluate pesticide spray drift are derived from studies of droplet size spectrum and spray drift field evaluations. These data contribute to the development of the overall exposure estimate and, along with data on toxicity for humans, fish and wildlife, or plants, are used to assess the potential hazard of pesticides to these organisms. A purpose common to all these tests is to provide data which will be used to determine the need for (and appropriate wording for) precautionary labeling to minimize the potential adverse effect to nontarget organisms.

h) *Environmental fate—*

(1) *General.* The data generated by environmental fate studies are used to: Assess the toxicity to man through exposure of humans to pesticide residues remaining after application, either upon reentering treated areas or from consuming inadvertantly-contaminated food; assess the presence of widely distributed and persistent pesticides in the environment which may result in loss of usable land, surface water, ground water, and wildlife resources; and, assess the potential environmental exposure of other nontarget organisms, such as fish and wildlife, to pesticides. Another specific purpose of the environmental fate data requirements is to help applicants and the Agency estimate expected environmental concentrations of pesticides in specific habitats where threatened or endangered species or other wildlife populations at risk are found.

(2) *Degradation studies.* The data from hydrolysis and photolysis studies are used to determine the rate of pesticide degradation and to identify pesticides that may adversely affect nontarget organisms.

(3) *Mobility studies.* These data requirements pertain to leaching, adsorption/desorption, and volatility of pesticides. They provide information on the mode of transport and eventual destination of the pesticide in the environment. This information is used to assess potential environmental hazards related to: Contamination of human and animal food; loss of usable land and water resources to man through contamination of water (including ground water); and habitat loss of wildlife resulting from pesticide residue movement or transport in the environment.

* * *

(4) *Accumulation studies.* Accumulation studies indicate pesticide residue levels in food supplies that originate from wild sources or from rotational crops. Rotational crop studies are necessary to establish realistic crop rotation restrictions and to determine if tolerances may be

needed for residues on rotational crops. Data from irrigated crop studies are used to determine the amount of pesticide residues that could be taken up by representative crops irrigated with water containing pesticide residues. These studies allow the Agency to establish label restrictions regarding application of pesticides on sites where the residues can be taken up by irrigated crops. These data also provide information that aids the Agency in establishing any corresponding tolerances that would be needed for residues on such crops. Data from pesticide accumulation studies in fish are used to establish label restrictions to prevent applications in certain sites so that there will be minimal residues entering edible fish or shellfish. These residue data are also used to determine if a tolerance or action level is needed for residues in aquatic animals eaten by humans.

* * *

Subpart G – Ecological Effects

§ 158.630 Terrestrial and aquatic nontarget organisms data requirements table.

a) *General.* Sections 158.100 through 158.130 describe how to use this table to determine the terrestrial and aquatic nontarget data requirements for a particular pesticide product. Notes that apply to an individual test including specific conditions, qualifications, or exceptions to the designated test are listed in paragraph (e) of this section.

b) *Use patterns.*

(1) The terrestrial use pattern includes products classified under the general use patterns of terrestrial food crop, terrestrial feed crop, and terrestrial nonfood crop. The aquatic use pattern includes products classified under the general use patterns of aquatic food crop and aquatic nonfood use patterns. The greenhouse use pattern includes products classified under the general use patterns of greenhouse food crop and greenhouse nonfood crop. The indoor use pattern includes products classified under the general use patterns of indoor food and indoor nonfood use.

(2) Data are also required for the general use patterns of forestry and residential outdoor use.

(3) In general, for all outdoor end-uses, including turf, the following studies are required: Two avian oral LD_{50}, two avian dietary LC_{50}, two avian reproduction studies, two freshwater fish LC_{50}, one freshwater invertebrate EC_{50}, one honeybee acute contact LD_{50}, one freshwater fish early-life stage, one freshwater invertebrate life cycle, and three estuarine acute LC_{50}/EC_{50} studies—fish, mollusk and invertebrate. All other outdoor residential uses, *i.e.*, gardens and ornamental will not usually require the freshwater fish early-life stage, the freshwater invertebrate life-cycle, and the acute estuarine tests.

Appendix A

c) *Key.* R = Required; CR = Conditionally required; NR = Not required; TGAI = Technical grade of the active ingredient; TEP = Typical end-use product; PAI = Pure active ingredient; EP = end-use product. Commas between the test substances (*i.e.*, TGAI, TEP) indicate that data may be required on the TGAI or the TEP depending on the conditions set forth in the test note.

d) *Table.* The following table shows the data requirements for nontarget terrestrial and aquatic organism. The table notes are shown in paragraph (e) of this section.

e) *Test notes.* The following test notes apply to terrestrial and aquatic nontarget organisms data requirements in the table to paragraph (d) of this section:

> (1) Data using the TGAI are required to support all outdoor end-use product uses including, but not limited to turf. Data are generally not required to support end-use products in the form of a gas, a highly volatile liquid, a highly reactive solid, or a highly corrosive material.
>
> (2) For greenhouse and indoor end-use products, data using the TGAI are required to support manufacturing-use products to be reformulated into these same end-use products or to support end-use products when there is no registered manufacturing-use product. Avian acute oral data are not required for liquid formulations for greenhouse and indoor uses. The study is not required if there is no potential for environmental exposure.
>
> (3) Data are required on one passerine species and either one waterfowl species or one upland game bird species for terrestrial, aquatic, forestry, and residential outdoor uses. Data are preferred on waterfowl or upland game bird species for indoor and greenhouse uses.
>
> (4) Data are required on waterfowl and upland game bird species.
>
> (5) Tests are required based on the results of lower tier toxicology studies, such as the acute and subacute testing, intended use pattern, and environmental fate characteristics that indicate potential exposure.
>
> (6) Higher tier testing may be required for a specific use pattern when a refined risk assessment indicates a concern based on laboratory toxicity endpoints and refined exposure assessments.
>
> (7) Environmental chemistry methods used to generate data associated with this study must include results of a successful confirmatory method trial by an independent laboratory. Test standards and procedures for independent laboratory validation are available as addenda to the guideline for this test requirement.
>
> (8) Data are required on one coldwater fish and one warmwater fish for terrestrial, aquatic, forestry, and residential outdoor uses. For indoor and greenhouse uses, testing with only one of either fish species is required.

Terrestrial and Aquatic Nontarget Organism Data Requirements

Guideline Number	Data Requirement	Use Pattern						Test substance	Test Note No.
		Terrestrial	Aquatic	Forestry	Residential Outdoor	Greenhouse	Indoor		
Avian and Mammalian Testing									
850.2100	Avian oral toxicity	R	R	R	R	CR	CR	TGAI	1, 2, 3
850.2200	Avian dietary toxicity	R	R	R	R	NR	NR	TGAI	1, 4
850.2400	Wild mammal toxicity	CR	CR	CR	CR	NR	NR	TGAI	5
850.2300	Avian reproduction	R	R	R	R	NR	NR	TGAI	1, 4
850.2500	Simulated or actual field testing	CR	CR	CR	CR	NR	NR	TEP	6, 7
Aquatic Organisms Testing									
850.1075	Freshwater fish toxicity	R	R	R	R	CR	CR	TGAI, TEP	1, 2, 8, 9, 26
850.1010	Acute toxicity freshwater invertebrates	R	R	R	R	CR	CR	TGAI, TEP	1, 2, 9, 10, 26
850.1025 850.1035 850.1045 850.1055 850.1075	Acute toxicity estuarine and marine organisms	R	R	R	R	NR	NR	TGAI, TEP	1, 9, 11, 12, 26
850.1300	Aquatic invertebrate life cycle (freshwater)	R	R	R	R	NR	NR	TGAI	1, 10, 12
850.1350	Aquatic invertebrate life cycle (saltwater)	CR	CR	CR	CR	NR	NR	TGAI	12, 14, 15
850.1400	Fish early-life stage (freshwater)	R	R	R	R	NR	NR	TGAI	1, 12, 13
850.1400	Fish early-life stage (saltwater)	CR	CR	CR	CR	NR	NR	TGAI	12, 15, 16
850.1500	Fish life cycle	CR	CR	CR	CR	NR	NR	TGAI	17, 18

850.1710 850.1730 850.1850	Aquatic organisms bioavailability, biomagnification, toxicity	CR	CR	CR	CR	NR	NR	TGAI, PAI, degradate	19
850.1950	Simulated or actual field testing for aquatic organisms	CR	CR	CR	CR	NR	NR	TEP	7, 20
Sediment Testing									
850.1735	Whole sediment: acute freshwater invertebrates	CR	CR	CR	CR	NR	NR	TGAI	21
850.1740	Whole sediment: acute marine invertebrates	CR	CR	CR	CR	NR	NR	TGAI	21, 23
	Whole sediment: chronic invertebrates freshwater and marine	CR	CR	CR	CR	NR	NR	TGAI	22, 23
Insect Pollinator Testing									
850.3020	Honeybee acute contact toxicity	R	CR	R	R	NR	NR	TGAI	1
850.3030	Honey bee toxicity of residues on foliage	CR	CR	CR	CR	NR	NR	TEP	24
850.3040	Field testing for pollinators	CR	CR	CR	CR	NR	NR	TEP	25

(9) EP or TEP testing is required for any product which meets any of the following conditions:
    i. The end-use pesticide will be introduced directly into an aquatic environment (e.g., aquatic herbicides and mosquito larvicides) when used as directed.
    ii. The maximum expected environmental concentration (MEEC) or the estimated environmental concentration (EEC) in the aquatic environment is $\geq$one-half the LC_{50} or EC_{50} of the TGAI when the EP is used as directed.
    iii. An ingredient in the end-use formulation other than the active ingredient is expected to enhance the toxicity of the active ingredient or to cause toxicity to aquatic organisms.

(10) Data are required on one freshwater aquatic invertebrate species.

(11) Data are required on one estuarine/marine mollusk, one estuarine/marine invertebrate and one estuarine/marine fish species.

(12) Data are generally not required for outdoor residential uses, other than turf, unless data indicate that pesticide residues from the proposed use(s) can potentially enter waterways.

(13) Data are required on one freshwater fish species. If the test species is different from the two species used for the freshwater fish acute toxicity tests, a 96-hour LC_{50} on that species must also be provided.

(14) Data are required on one estuarine/marine invertebrate species.

(15) Data are required on estuarine/marine species if the product meets any of the following conditions:
    i. Intended for direct application to the estuarine or marine environment.
    ii. Expected to enter this environment in significant concentrations because of its expected use or mobility patterns.
    iii. If the acute LC_{50} or EC_{50} <1 milligram/liter (mg/l).
    iv. If the estimated environmental concentration (EEC) in water is $\geq$0.01 of the acute EC_{50} or LC_{50} or if any of the following conditions exist:
        A. Studies of other organisms indicate the reproductive physiology of fish and/or invertebrates may be affected.
        B. Physicochemical properties indicate bioaccumulation of the pesticide.
        C. The pesticide is persistent in water (e.g., half-life in water >4 days).

(16) Data are required on one estuarine/marine fish species.

(17) Data are required on estuarine/marine species if the product is intended for direct application to the estuarine or marine environment, or the product is expected to enter this environment in significant concentrations because of its expected use or mobility patterns.

(18) Data are required on freshwater species if the end-use product is intended to be applied directly to water, or is expected to be transported

Appendix A 165

to water from the intended use site, and when any of the following conditions apply:
  i. If the estimated environmental concentration (EEC) is ≥0.1 of the no-observed-effect level in the fish early-life stage or invertebrate life cycle test;
  ii. If studies of other organisms indicate that the reproductive physiology of fish may be affected.
(19) Not required when:
  i. The octanol/water partition coefficients of the pesticide and its major degradates are <1,000; or
  ii. There are no potential exposures to fish and other nontarget aquatic organisms; or
  iii. The hydrolytic half-life is <5 days at pH 5, 7 and 9.
(20) Data are required based on the results of lower tier studies such as acute and chronic aquatic organism testing, intended use pattern, and environmental fate characteristics that indicate significant potential exposure.
(21) Data are required if:
  i. The half-life of the pesticide in the sediment is ≤10 days in either the aerobic soil or aquatic metabolism studies and if any of the following conditions exist:
    A. The soil partition coefficient (Kd) is ≥50.
    B. The log Kow is ≥3.
    C. The Koc ≥1,000.
  ii. Registrants must consult with the Agency on appropriate test protocols prior to designing the study.
(22) Data are required if:
  i. The estimated environmental concentration (EEC) in sediment is >0.1 of the acute LC_{50}/EC_{50} values and
  ii. The half-life of the pesticide in the sediment is >10 days in either the aerobic soil or aquatic metabolism studies and if any of the following conditions exist:
    A. The soil partition coefficient (Kd) is ≥50.
    B. The log Kow is ≥3.
    C. The Koc ≥1,000.
  iii. Registrants must consult with the Agency on appropriate test protocols prior to designing the study.
(23) Sediment testing with estuarine/marine test species is required if the product is intended for direct application to the estuarine or marine environment or the product is expected to enter this environment in concentrations which the Agency believes to be significant, either by runoff or erosion, because of its expected use or mobility pattern.
(24) Data are required only when the formulation contains one or more active ingredients having an acute LD_{50} of <11 micrograms per bee as

determined in the honey bee acute contact study and the use pattern(s) indicate(s) that honey bees may be exposed to the pesticide.

(25) Required if any of the following conditions are met:

    i. Data from other sources (Experimental Use Permit program, university research, registrant submittals, etc.) indicate potential adverse effects on colonies, especially effects other than acute mortality (reproductive, behavioral, etc.);

    ii. Data from residual toxicity studies indicate extended residual toxicity.

    iii. Data derived from studies with terrestrial arthropods other than bees indicate potential chronic, reproductive or behavioral effects.

(26) The freshwater fish test species for the TEP testing is the most sensitive of the species tested with the TGAI. Freshwater invertebrate and acute estuarine and marine organisms must also be tested with the EP or TEP using the same species tested with the TGAI.

* * *

Subpart L – Spray Drift

§ 158.1100 Spray drift data requirements table.

a) *General.* Sections 158.100 through 158.130 describe how to use this table to determine the spray drift data requirements for a particular pesticide product. Notes that apply to an individual test, including specific conditions, qualifications, or exceptions to the designated test are listed in paragraph (e) of this section.

b) *Use patterns.* The terrestrial use pattern includes products classified under the general use patterns of terrestrial food crop and terrestrial nonfood crop. The aquatic use pattern includes products classified under the general use patterns of aquatic food crop and aquatic nonfood. The greenhouse use pattern includes products classified under the general use patterns of greenhouse food crop and greenhouse nonfood crop. Data are also required for the general use patterns of forestry use, residential outdoor use, and indoor use.

c) *Key.* CR = Conditionally required; NR = Not required; TEP = Typical end-use product; MP = Manufacturing use product; EP = End-use product.

d) *Table.* The following table lists the data requirements that pertain to spray drift. The table notes are shown in paragraph (e) of this section.

* * *

Subpart N – Environmental Fate

§ 158.1300 Environmental fate data requirements table.

a) *General.* All environmental fate data, as described in paragraph (c) of this section, must be submitted to support a request for registration.

b) *Use patterns.*

Appendix A

(1) The terrestrial use pattern includes products classified under the general use patterns of terrestrial food crop, terrestrial feed crop, and terrestrial nonfood. The aquatic use pattern includes the general use patterns of aquatic food crop, and aquatic nonfood. The greenhouse use pattern includes both food and nonfood uses. The indoor use pattern includes food, nonfood, and residential indoor uses.

(2) Data are also required for the general use patterns of forestry use and residential outdoor use.

c) *Key.* CR = Conditionally required; NR = Not required; R = Required; PAIRA = Pure active ingredient radio-labeled; TGAI = Technical grade of the active ingredient; TEP = Typical end-use product.

d) *Table.* The following table shows the data requirements for environmental fate. The test notes are shown in paragraph (e) of this section.

e) *Test notes.* The following test notes apply to the requirements in the table to paragraph (d) of this section:

(1) Study is required for indoor uses in cases where environmental exposure is likely to occur. Such sites include, but are not limited to, agricultural premises, in or around farm buildings, barnyards, and beehives.

(2) Not required when the electronic absorption spectra, measured at pHs 5, 7, and 9, of the chemical and its hydrolytic products, if any, show no absorption or tailing between 290 and 800 nm.

(3) Not required when the chemical is to be applied only by soil injection or is incorporated in the soil.

(4) Requirement based on use patterns and other pertinent factors including, but not limited to, the Henry's Law Constant of the chemical. In view of methodological difficulties with the study of photodegradation in air, prior consultation with the Agency regarding the protocol is recommended before the test is performed.

(5) Required for aquatic food and nonfood crop uses for aquatic sites that are intermittently dry. Such sites include, but are not limited to, cranberry bogs and rice paddies.

(6) Adsorption and desorption using a batch equilibrium method is preferred. However in some cases, for example, where the pesticide degrades rapidly, soil column leaching with unaged or aged columns may be more appropriate to fully characterize the potential mobility of the parent compound and major transformation products.

(7) Environmental chemistry methods used to generate data associated with this study must include results of a successful confirmatory method trial by an independent laboratory. Test standards and procedures for independent laboratory validation are available as addenda to the guideline for this test requirement.

(8) Requirement for terrestrial uses is based on potential for aquatic exposure and if pesticide residues have the potential for persistence, mobility, nontarget aquatic toxicity or bioaccumulation. Not required for aquatic residential uses. Field testing under the terrestrial field dissipa-

tion requirement may be more appropriate for some aquatic food crops, such as rice and cranberry uses, that are managed to have a dry-land period for production. The registrant is encouraged to consult with the Agency on protocols.

(9) Agency approval of a protocol is necessary prior to initiation of the study.

(10) This study may be triggered if there is specific evidence that the presence of one pesticide can affect the dissipation characteristics of another pesticide when applied simultaneously or serially.

(11) Required if the weight-of-evidence indicates that the pesticide and/or its degradates is likely to leach to ground water, taking into account other factors such as the toxicity of the chemicals(s), available monitoring data, and the vulnerability of ground water resources in the pesticide use area.

(12) If the terrestrial dissipation study cannot assess all of the major routes of dissipation, the forestry.

Table—Spray Drift Data Requirements

Guideline Number	Data Requirement	Use Pattern									Test substance		
		Terrestrial		Aquatic		Greenhouse							
		Food Crop	Nonfood Crop	Food	Nonfood	Food Crop	Nonfood Crop	Forestry	Residential Outdoor	Indoor	MP	EP	Test Note No.
201–1	Droplet size spectrum	CR	CR	CR	CR	NR	NR	CR	NR	NR	TEP	TEP	1
202–1	Droplet size spectrum	CR	CR	CR	CR	NR	NR	CR	NR	NR	TEP	TEP	1

e) *Test notes.* The following notes apply to the requirements in the table to paragraph (d) of this section:

(1) This study is required when aerial applications (rotary and fixed winged) and mist blower or other methods of ground application are proposed and it is estimated that the detrimental effect level of those nontarget organisms expected to be present would be exceeded. The nontarget organisms include humans, domestic animals, fish and wildlife, and nontarget plants.

Table—Environmental Fate Data Requirements

Guideline Number	Data Requirement	Use Pattern						Test substance	Test Note No.
		Terrestrial	Aquatic	Greenhouse	Indoor	Forestry	Residential Outdoor		
Degradation Studies - Laboratory									
835.2120	Hydrolysis	R	R	R	CR	R	R	TGAI or PAIRA	1
835.2240	Photodegradation in water	R	R	NR	NR	R	NR	TGAI or PAIRA	2
835.2410	Photodegradation on soil	R	NR	NR	NR	R	NR	TGAI or PAIRA	3
835.2370	Photodegradation in air	CR	NR	CR	NR	CR	CR	TGAI or PAIRA	4
Metabolism Studies - Laboratory									
835.4100	Aerobic soil	R	CR	R	NR	R	R	TGAI or PAIRA	5
835.4200	Anaerobic soil	R	NR	NR	NR	NR	NR	TGAI or PAIRA	--
835.4300	Aerobic aquatic	R	R	NR	NR	R	NR	TGAI or PAIRA	--
835.4400	Anaerobic aquatic	R	R	NR	NR	R	NR	TGAI or PAIRA	--
Mobility Studies									
835.1230 835.1240	Leaching and adsorption/desorption	R	R	R	NR	R	R	TGAI or PAIRA	6
835.1410	Volatility - laboratory	CR	NR	CR	NR	NR	NR	TEP	4
835.8100	Volatility - field	CR	NR	CR	NR	NR	NR	TEP	--
Dissipation Studies - Field									
835.6100	Terrestrial	R	CR	NR	NR	CR	R	TEP	5, 7, 12
835.6200	Aquatic (sediment)	CR	R	NR	NR	NR	NR	TEP	7, 8
835.6300	Forestry	NR	NR	NR	NR	CR	NR	TEP	7, 9, 12
835.6400	Combination and tank mixes	CR	CR			NR	NR	TEP	10
Ground Water Monitoring									
835.7100	Ground water monitoring	CR	NR	NR	NR	CR	CR	TEP	7, 9, 11

Appendix B

Biographical Information on the Committee on Ecological Risk Assessment under FIFRA and ESA

Judith E. McDowell (*Chair*) is a senior scientist and former Biology Department chair at Woods Hole Oceanographic Institution. Her research interests include physiological ecology of marine animals, developmental and energetic strategies of marine animals, physiological effects of pollutants on marine animals, and invertebrate nutrition. She has served on several National Research Council committees, including the Committee on Oil in the Sea: Phase I—Update of Inputs and the Committee on Research and Peer Review in EPA. Dr. McDowell earned a PhD in zoology from the University of New Hampshire.

H. Resit Akcakaya is a professor in the Department of Ecology and Evolution at Stony Brook University. His research focuses on methods and approaches for assessing the vulnerability of species to extinction, evaluating the effects of landscape dynamics on species persistence, projecting human land use on the basis of human population trends, and predicting the vulnerability of species to global climate change. He worked as a senior scientist at Applied Biomathematics, where he was one of the principal architects of the RAMAS library of software and developed models for risk assessment and modeling of metapopulations, for integrating metapopulation dynamics with geographic information systems, and for incorporating uncertainty into International Union for Conservation of Nature (IUCN) criteria for threatened species. Dr. Akcakaya has also been involved in practical and theoretical research on problems of species conservation, including several population-viability analysis studies. He has over 100 publications in conservation biology and theoretical ecology, including four books, and is a coauthor of two widely used textbooks (*Risk Assessment in Conservation Biology* and *Applied Population Ecology*). In addition, Dr. Akcakaya serves on the editorial boards of *Conservation Biology* and *Population Ecology* and is chair of the IUCN Red List Standards and Petitions Subcommittee. Dr.

Akcakaya earned a PhD in ecology and evolution from the State University of New York at Stony Brook.

Mary Jane Angelo is professor of law and director of the Environmental and Land Use Law Program at the University of Florida's Levin College of Law. Her research focuses on environmental law, water law, agricultural law, pesticide law, endangered species law, biotechnology law, and the integration of law and science. Before joining the faculty, Ms. Angelo served as an attorney in the US Environmental Protection Agency's Office of General Counsel and as senior assistant general counsel for the St. Johns River Water Management District. In addition, she has served on the National Research Council Committee on Independent Scientific Review of Everglades Restoration Progress. Ms. Angelo earned an MS in entomology and JD from the University of Florida.

Patrick Durkin is cofounder and principal scientist of Syracuse Environmental Research Associates, a small business engaged in chemical and biological risk assessment and documentation. He has been responsible for developing safety evaluations for chemical and biological agents on the basis of a synthesis of toxicological data, environmental persistence, and exposure estimates. Dr. Durkin has conducted numerous risk assessments and risk assessment method development tasks for the US Department of Agriculture, the US Environmental Protection Agency, and the Centers for Disease Control and Prevention Agency for Toxic Substances and Disease Registry. Dr. Durkin earned a PhD in environmental and forest zoology from SUNY College of Environmental Science and Forestry.

Erica Fleishman is a researcher in the John Muir Institute of the Environment at the University of California, Davis. Her research focuses on integration of conservation science with management and policy, especially in the intermountain western United States and California. Her work focuses on predictive modeling of occupancy and faunal responses to changes in climate, land cover, land use, and connectivity. Dr. Fleishman is a coauthor of curricula in applications of remote sensing to environmental sciences and ecological modeling. She has convened multidisciplinary teams to analyze and synthesize concepts and data on diverse topics and has facilitated or advised on the science process for multiple habitat conservation plans and natural community conservation plans in California. Dr. Fleishman is past editor in chief of *Conservation Biology* and serves on the editorial boards of *Global Ecology, Biogeography,* and *Ecography.* Dr. Fleishman earned a PhD in ecology, evolution, and conservation biology from the University of Nevada, Reno.

Anne Fairbrother is a principal scientist for Exponent's ecosciences practice. She has more than 30 years of experience in ecotoxicology, wildlife toxicology, contaminated-site assessment, and regulatory science for existing and emerging chemicals in the United States and Europe. Dr. Fairbrother has participated in or

Appendix B

led the development of guidance documents for ecological risk assessments, such as the US Environmental Protection Agency's (EPA) *Framework for Metals Risk Assessment*, the British Columbia Ministry of Environment's guidance for implementing Tier 1 ecological risk assessments of contaminated sites, and EPA's ecological soil screening levels. Recently, she served on a science advisory panel for the state of Utah and as a consultant to the British Columbia Ministry of Environment to set site-specific water-quality standards for selenium that protect fish and wildlife. Dr. Fairbrother has served as president of the Society of Environmental Toxicology and Chemistry, the American Association of Wildlife Veterinarians, and the Wildlife Disease Association. In addition, she has been a member of the National Research Council's Committee on Animals as Monitors of Environmental Hazards. Dr. Fairbrother earned a DVM from the University of California, Davis and a PhD in veterinary science from the University of Wisconsin-Madison.

Daniel Goodman was a professor of ecology at Montana State University. His research interests included environmental statistics, risk analysis, population dynamics, and environmental modeling. Dr. Goodman was a member of the Silvery Minnow PVA Working Group (Middle Rio Grande Endangered Species Collaborative), the Fish Passage Center Oversight Board of the Northwest Power and Conservation Council, the Hawaiian Monk Seal Recovery Team, and the Cook Inlet Beluga Whale Recovery Team. Dr. Goodman earned a PhD in zoology from Ohio State University.

William L. Graf is University Foundation Distinguished Professor Emeritus of the Department of Geography of the University of South Carolina and Regents Professor Emeritus in Geography at Arizona State University. His research interests include fluvial geomorphology and hydrology and policy for public land and water with an emphasis on river channel and habitat change, human effects on rivers, contaminant transport and storage in rivers, and the downstream effects of large dams. He has served as a science-policy adviser on more than 40 committees for federal, state, and local agencies and organizations. In addition, Dr. Graf has chaired and been a member of many National Research Council committees, including those focused on the Klamath River, the Platte River, the Everglades, the Missouri River, and watershed management. He is chair of the NRC Geographical Sciences Committee, a national associate of the National Academy of Sciences, and a fellow of the American Association for the Advancement of Science. Dr. Graf earned his PhD in physical geography from the University of Wisconsin-Madison with a certificate in water-resources management.

Philip M. Gschwend is a professor of civil and environmental engineering at the Massachusetts Institute of Technology. His research interests are environmental organic chemistry, phase exchanges and transformation processes, modeling fates of organic pollutants, roles of colloids and black carbons, and passive sampling for site evaluation. The overall objective of his research is to develop means of predicting the fate of organic chemicals in natural and engineered environments. His research includes the study of such processes as sorption, air-water exchange, and biodegradation. In addition, Dr. Gschwend conducts field observations in water and sediments of groundwater, lakes, estuaries, and the ocean to validate the predictions. Dr. Gschwend earned a PhD in geochemistry from the Woods Hole Oceanographic Institution.

Bruce K. Hope is a principal environmental scientist with CH2M HILL. His expertise includes preparation and review of human, ecological, and probabilistic risk assessments; exposure modeling; development of air-toxics benchmarks; identification and management of persistent and bioaccumulative chemicals; and evaluation and communication of health and environmental risks associated with chemical releases. Dr. Hope has served on a number of US Environmental Protection Agency (EPA) Science Advisory Board committees, including that on Ecological Risk Assessment—An Evaluation of the State-of-the-Practice and EPA's Regulatory Environmental Modeling Guidance Advisory Panel. In addition, he was a member of the National Research Council Committee on Improving Risk Analysis Approaches Used by the U.S. EPA. Dr. Hope earned a PhD in biology from the University of Southern California.

Gerald A. LeBlanc is the head of and a professor in the Department of Environmental and Molecular Toxicology of North Carolina State University. His research interests include environmental signaling, sex determination and differentiation, and toxicity assessment of chemical mixtures. Dr. LeBlanc has been a member of the Executive Committee of the Research Triangle Environmental Health Collaborative, of the FIFRA National Science Advisory Panel on the potential for atrazine to affect amphibian gonadal development, and of the National Institute of Environmental Health Sciences Expert Panel on Hazards of Bisphenol A to Humans and the Environment. Dr. LeBlanc earned a PhD in biology from the University of South Florida.

Thomas P. Quinn is a professor in the School of Aquatic and Fishery Sciences of the University of Washington. His research interests focus on the behavior, ecology, evolution, and conservation of salmon, trout, and related fishes. Dr. Quinn's research blends a variety of approaches, including tagging, telemetry, direct observations, and laboratory experiments. He is studying the patterns of spawning-site selection and reproductive behavior of salmon, movements and migration patterns, evolutionary adaptations of salmon to their environments, and predator-prey ecology. He has served on the National Research Council Committee on Protection and Management of Pacific Northwest Anadromous Salmonids. Dr. Quinn earned a PhD in fisheries from the University of Washington.

Appendix B

Nu-May Ruby Reed recently retired as a staff toxicologist with the California Environmental Protection Agency (Cal/EPA) Department of Pesticide Regulation, where she was the lead scientist on risk-assessment issues. Her research interests were evaluating health risks posed by and developing risk-assessment guidelines on pesticides. She has been on several Cal/EPA working groups that initiate, research, and revise risk-assessment guidelines and policies, and she represented her department in task forces on community concerns and emergency response, risk-management guidance, and public education. Dr. Reed has been a member of several National Research Council Committees, including the Committee on Risk Analysis and Reviews, and is a current member of the Committee on Acute Exposure Guideline Levels. Dr. Reed earned a PhD in plant physiology from the University of California, Davis.